"IS THIS THING ON?"

Collected and compiled by

ROBIN MARSHALL

Illustrated by

Dylan Bluver

FIRST EDITION

MCA BOOKS

Marshall Arts Communications

First Edition, 2010, manufactured in USA
1 2 3 4 5 6 7 8 9 10

All illustrations in this book are by the kind permission of Dylan Bauver.

Photographs are by the kind permission of the owner, with all right reserved by the authors.

All text underneath listed authors are the copyright of the author and are printed here with kind permission. All words and statements below listed authors are those of the individual author and not those of Robin Marshall, apart from comments in bold italic.

ISBN: 978-0-615-37137-5

DEDICATION

This book is dedicated to my children. I'm hoping that they've seen the hard work and dedication that was put in by myself, my son and all of these radio author's, for the sole purpose of the betterment of other children not as fortunate. We should live each day as if it is our *last* day, not just selfishly, but to the degree of giving as well. A little bit of 'give'...goes a LONG way!

Love and hugs to you all!

Robin

CONTENTS

INTRODUCTION

Hello my wonderful friends!

After editing, which meant, I had to READ every story… I'm lucky I don't have nightmares every night! We are truly, the most incredible breed. To have survived all of this, and to have lived to laugh about it? Holy Moley!

The stories keep coming in, so I wanted to mention that this will be "Book 1." If you missed the cut off, …welcome to "Book 2." Just send me an email with the following:

1. First radio story
2. Your worst radio experiences
3. Where you are today.

I think we needed this! In turbulent times, it's always good to look back, and see where we came from, the trials and tribulations of what made us "who we are today," and be able to realize, that just because we may have recently lost our gigs, it wasn't due to "lack of talent." "Downsizing happens." It can NOT erase 'who we were', and more importantly, 'who we ARE'.

The illustrations in this book are done by my son, Dylan. He's 18, getting ready for college and is fully equipped with Photoshop. He's available and eager to do graphic designing for you. Logos, projects…, whatever you may need. I highly recommend him! This was my way of introducing him to our world, and you to his art. Nothing like 'promoting from within!'

He is about to begin his first semester at Iona, of which he was given a full ride for his classes based upon his "goalie skills! Yup!" "An Artist,in an Athlete's body!" (spoken like the non-biased Mom that I am!)

You'll notice occasionally throughout the book **(my personal notes in *bold italics*).** Some were just IMPOSSIBLE not to respond to! OMG. I hope you also notice, there was no "set amount of space per person." This was your opportunity to shine, and as long as it was compelling? I let ya' rip!

Much love and adoration to all of you, and if you'd like to be in

'Book 2," my email is Robin.MarshallsOffice@gmail.com .

I felt it was appropriate to start with my Dad's story... and finish with mine!

Robin
Xoxoxoxoxo

Perry Marshall

When you do an all night talk show as I did for many years on KDKA, you run the gamut of topics. They'd range from politics to war, to unfaithful spouses, to bratty kids that wouldn't listen, and even aches and pains.

There have been some scary moments, for example the night when a guy called my show, and in a very threatening tone of voice said, "I know where your car is parked." He then went on to describe exactly where that was. He added, "but you won't get there alive." I told my producer to call the police, and they gave me an armed escort at 5am.

Then there was the night when I was tired of the 'heavy topics,"so I decided to invite listeners to join me in playing the 'nose guitar.' How do you do that? You squeeze one nostril closed with one finger, hum in a high tone, and strum the other nostril with your other finger. That was at 4 am. The following night, I got a call from a woman in Nebraska who's husband woke up and said, "What the hell are you doing?" She told him, "I'm playing the nose guitar!" He then threatened to divorce her. ***(RM-as***

did my mother!)

Every now and then I'd get a call from a threatened suicide, usually a drunk. I'd listen for a while, and then recommend the suicide hotline. However, at 3am one night, the caller seemed serious. He said that he had a gun pointed at his head, and was going to 'blow his brains out." I kept him on the line, while signaling my producer to pay close attention. The caller was sobbing. He said that he'd lost his job as an iron worker, and that was the last straw. He told me that he had a lot of life insurance, and that his family would be better off without him. I tried to console him by asking him, if he was sure this was the right thing to do. He kept sobbing and said that he was sure. Then, remembering my past life as an insurance agent, I told him that some policies don't pay for suicide.

At that moment, I heard a gunshot followed by a scream. I heard a door open, and a woman's voice said, "Who is this?" I told her who I was, and asked if he was alive. She said, "He shot himself... in the arm." I told her to get rid of the gun, and had my producer call the police.

I thought that was the end of it....BUT! The station got a call from Western Psychiatric Hospital. They wanted to find out if talk shows had any therapeutic value.

I found myself in a large auditorium on a very large stage, with a psychiatrist on either side of me. I said, "No shock treatment please." Well, everyone started laughing and then they got serious. They wanted to know what credentials I had, what kind of education, what degrees, etc. I told them that I didn't have much formal education, no degree of any kind, but I did have a lot of one on one experience.

Today, I'm retired, I enjoy every sunrise and sunset from my balcony. I still do the NY Times Cross word puzzle every day, ***(RM-a little known fact here, he used to do the puzzle while he was on the air- every night- and he'd FINISH it!)*** I belong to library group of actors, and occasionally when in NY, will record for my daughter Robin. I'm extremely proud of her accomplishments. Aren't you?

Chris Roth

Anyone had "the dream"?? I've had the one...and it always centers around being at WRQX...I go out into the hallway and start talking to

Chris Core from WMAL (yeah, odd dream as it is...) - and the song starts fading. I start to go back into the studio and either....A - the door is locked, B - nothing happens when I start pushing buttons or C - my favorite - the console is gone and replaced by a single button. Gotta quit eating before bed... ***(RM-as IF that's why!)***

AND…

Nothing like being on the air when you are 16...and the studio is at a transmitter site and all of a sudden you hear dead air, even though the cart is still playing. You start the next cart...dead air. Transmitter is on the air... Call the engineer... he comes and informs me that a truck has knocked down tower two. Oops...Wow...we might have to start chapters in here, Robin... ***(RM-notice he's at the beginning of the book! What foresight you had Chris!)***

Tom Brammann

Our company had two separate stations. One was an FM adult contemporary station and the other was an AM country station. I was getting ready to start my shift on the FM station when the PM called and told me that the AM talent was sick and that I had to do both air shifts at the same time! Fortunately the studios were next door to each other. After some quick timing planning I did the parallel airshifts with very few noticeable flubs.

John Dowling

One version of my dead air dream: I suddenly realize I have less than a minute to news, run to the wire machine and tear off about 5 feet of copy, then race back to the studio. (Of course, this happened plenty of times in real life.) Then, as the record ends I scan down the copy, and there is nothing but ag reports and partial ball scores.

Joel Denver

Robin ...

Thanks for this opportunity!

Here we go ...

I got bitten by the radio bug early, and being persistent achieved my start in radio at age 15 at WINX-A/Rockville, MD in January, 1968. At 1600kHz, we had a 1000w non- directional signal daytime and a three tower direction array at night -- and if the wind was blowing right, we made it across the Maryland/DC border -- it was one of the worst signals in DC, but it was my first gig.

WINX was located in an old two-story house overlooking busy Rockville Pike and Hungerford Rd -- and a graveyard. Studios were on the second floor -- and if you were walking driving by, you could see into the studio. Not always a good thing!

One Saturday night, I was "on the can" during a 5 minute Mutual News feed -- when 30 seconds into "my mission", the newscast suddenly ends, and the network tone comes back up. I run down the hall, pants around my ankles, face plant twice before getting to the studio.

The control board is on a riser -- making it easy to see ... everything! Couldn't sit down -- and there I am "entertaining" drivers as I scramble to que a song and kill the network tone -- and the hotline is ringing -- it's the GM who wants to know "what's going on?"

When I first started and did weekend mornings, the first thing I had to do was a 15 minute newscast -- it was a rip n' read from the AP or UPI Newswire, which was all piled up from being unattended overnight. There were a lot of "hard to pronounce" Vietnamese and Russian names and towns that were in the news, as the Vietnamese War lead the news, and the Cold War with Russia was in full swing. Rather than try and struggle through these words -- a few seconds ahead of reading them, I'd close the mic mid-word ... keep my pacing and then open the mic a few seconds later mid-word ... as if we'd dropped the signal for a few seconds! Worked everytime!

One of those horrible remote broadcast stories involved being at WNOR-A (THE BIG 1230)/NORFOLK. We had a homemade studio "The Mini Mobile Music Maker" and it was cooled by a window air conditioner that shut off when you keyed the mic -- needless to say our breaks were short. When we used to cover the surfing championships, we had to set up on the beach, no shade, and were constantly moving "wire copy" across the glass as the sun moved -- as the 45s would actually begin to warp as they were playing in the direct sunlight -- which was made even worse -- when the mic was on.

All totalled, I spent 13 years in radio at WINX-A, then WPGC-A&M/ WASHINGTON, WNOR-A/NORFOLK, WAYS-A/CHARLOTTE (where a redneck shot out the guy wire insulators on Tower 1 causing it to lean out of plumb), WFIL-A, then went on to be PD at WMJX (96X)/MIAMI (where I saved one of my DJs from being taken into custody during a domestic dispute that spilled over onto station property), moved to SAN DIEGO and programmed KCBQ-A, then onto the PD chair at KSLQ (now KYKY)/ ST. LOUIS, then my final radio stop as PD at WBSB (B104)/BALTIMORE. ***(RM- Jesus Christ! Now we know why you prefer the WRITTEN word, over the "spoken word!")***

One of my most fun times in radio was at WFIL-A/PHILADELPHIA. I was 21 and was doing late nights and became MD. When John Lennon came to town to help us during a charity weekend I got to hang out with him a lot. He was no longer with Yoko by this point, so we talked a ton, I interviewed him on the air live. We even partied together, and yes, I hooked him up with one of our station's favorite "hitline honeys" for the weekend. John got on the train back to New York very tired, and very happy!

From there, I joined RADIO & RECORDS in SEPTEMBER of 1980 as the Assistant Editor in the DC Office, then moving to LOS ANGELES as the CHR Editor for 14 years -- leaving at the end of 1994.

In JANUARY 1995, my wife RIA and I founded the ALL ACCESS MUSIC GROUP,INC, and we went online with AllAccess.com in OCTOBER 1995. We are now a staff of 30 creative and energized people who are committed to providing the radio and music industries with the most up to date and accurate news and music information, on demand. With offices headquartered in MALIBU, CA, and NASHVILLE, TN with a two-person presence in NEW YORK CITY, AllAccess.com offers format sections in Top 40/Mainstream, Top 40/Rhythm-Crossover, Urban, Hot/Modern/AC, Alternative, Rock, Triple A, Country, and Contemporary Christian formats, and is poised for the future.

Joel Denver
President/Publisher
AllAccess.com

Tom Brehmer

I've had the nightmare, yes, but I had a dream recently that I was on the air doing everything perfectly. Had funny jokes, great tunes and listeners burning up the phone lines. Then I woke up and remembered, I've been off the air for over three years!

Marybeth Rockwell Walleisa

I'm sure that many of us have moved from one station to the next but, at the new station, given out either the slogan or the phone number for the old station. Of course, it happened to me when I had no time left on my song intro to correct myself! In that situation, all you can do is laugh and hope your boss has a sense of humor. Thankfully, mine did! :-)

Landon L. Horne

I first and foremost want to THANK YOU for being my FB friend. I also noticed you were writing a book about us DJ's. I would like to be mentioned in your book.

I would like to tell you about us "Rollover DJ's" (you can research what I tell you if you like). We are only known by our FCC license.

We were hired under the table for little to no pay. We were hired to save the radio station's ass. Hired so that the GM's can stay within their budgets. So no on air talents would have to work overtime, saving money. All we ever wanted was to play the music. We never knew the "right" people at the stations. We filled a void of airtime. We say what we wanted and played what we wanted (the Last D.J-Tom Petty) and "the Big Man" just can't take it.

We need to be recorded (documented) somewhere. Because since we were hired 'under the table" we were never documented.

Well that's about all I have I have on my dissertation.
Please give me and all the other "unsung heroes" of radio, that saved the radio station's ass, us "Rollover" DJ's some brief mention in your book.

Like I said earlier, feel free to research what I am saying.

Well I am sorry if I sound like I am on a "soapbox," but all I ever wanted was to be on the radio. ***(RM-Landon, most of us on the air, applaud the "likes of you!")*** The stations I worked at were, WJKI, WKDY, and WCKN, on air as Landon Lee.

Thanks, and friends always, Landon. (aka. "Big Dog!")

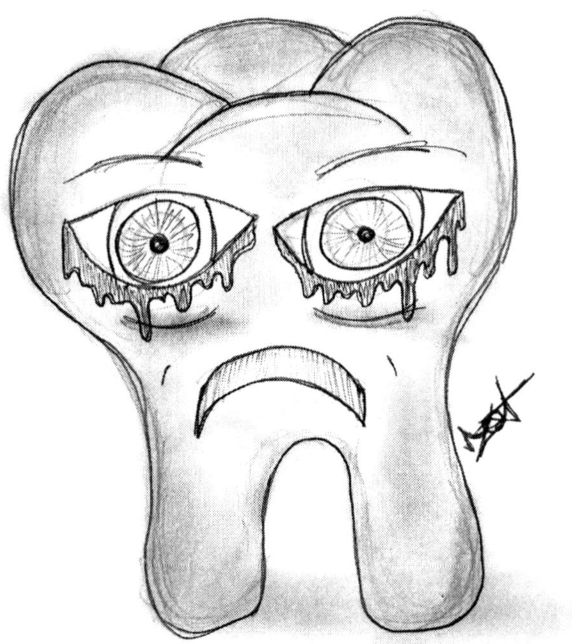

Greg Austin

When I was 16 I got my first gig working as a part-timer for KWBB-AM Wichita, KS. I worked Fri. nite, Sat. nite, and Sunday morning. The first Sunday morning I worked I had my wisdom teeth pulled that Friday. I made the 6 am show but about 7.30 am I fell asleep on the board because of the pain killers I was taking. The next thing I knew the weekday morning man, Jeff Hilldabrant was shaking my shoulder telling me that the board op at the church was on line one asking about the live feed.

Surprisingly they kept me around for a year and a half until I got a gig as a cameraman for the CBS station. They were not happy to see me leave as they couldn't get anyone else to work the hours they were having me do. That was 37 years ago.

Bruce Kettelle

Here is my detailed college radio story:

I have a hard time keeping it clean over the phone. In the spring of 1975 I dj'd at WNEC-FM, New England College Radio. Our chief announcer Steven D. Goodwin did a party show every Friday night. He had a regular segment called, "Check In" for callers to share where they were partying that evening. On this particular Friday I was at the college pub and was one of the first callers. Before I realized it I had said on live air "and the pub is fucking packed!" I paused briefly as I realized what had come out and then blurted "Oh no I said fuck on the air" quickly followed by "Oh shit I said fuck again," at which point I started banging the phone on top of the nearby cigarette machine, which reverberated loudly and started accusing a passerby of cursing into the phone and quickly hung up. ***(RM- THIS tops ALL!)***

After college I worked as a producer and the music assistant at WBCN-FM without much incident. I have not worked in radio since the early 1980's and currently operate several web sites while growing corn and soybeans on the family farm in Ohio.

Dan Hutti

I am sure you all know about getting a gig just because you were there at the right time. I actually got two of them:

Old story, guy ran out to smoke a joint and forgot the studio keys. Record just went shhhh.boomp, shhhh-boompet all for at least 45 minutes until someone came with a key.

Next day... he got fired and ...I was just there at the right time. Guy saw a ghost so he quit during his shift and left... he called the PD from down the hill the station was on.... and again I was just there at the right time. ***(RM- not only right place, right time- FIRST time I've ever heard this situation arising!)***

Nikki Landry

My first radio gig: KYQQ in Wichita, KS. No longer exists. I drove down there and asked to speak to the "manager" and out came Kevin Crash Davis. I asked him if he was hiring for DJ's. His reply? "And... you're who?" I gave him my name. He then asked me, "do you have any experience in radio?" I replied "You need EXPIERENCE to be a DJ? ...all you do is talk?" ***(RM-I giggle every time I read this Nikki!)*** He then asked me if I had time to do an "air check," so I said, "Sure!" I thought he wanted to check my breathing... so he puts me in a room and set it up for me to record a commercial. I did, and he said, "leave your name and number and if I'm interested I'll call you"...5 days later he called me... and the rest as they say, "Is History."

My worst radio nightmare: I always had a habit of saying "Uh" before I went on the air, so I had a nightmare one night that I said it on the air and that the radio kept skipping so all you heard was me saying, "Uh, uh, uh, uh, uh."

Today, after moving here from doing a mid-day show full time in San Francisco, I'm back in the Nations Capitol in Washington, DC doing PM drive on 94.7 Fresh FM.

Nikki Landry-Vice President of Relevance

Bob Heck

Guess I finally have one. I'm all hopped up on cough medicine and sinus stuff and I'm sucking down coffee like crazy (probably not the best combo), so I start my break and I fell some gas building up in my chest so I hit the bed for the next feature which lasts about 20 seconds and let out the biggest most blatant belch ever. To my horror, I realize I never shut my mic off as I usually do. I've since taken about 30 calls filled with hysterical laughter. No one offended thankfully.

Barry Michaels

My first on air job was at WYTI Radio in Rocky Mount, Virginia. You might be interested in knowing that our slogan was 'The Mighty Whitey'. 1kw daytimer... I spoke mostly to cows. ***(RM-So humble!)***

My worst experience?

One of our personalities at a small North Carolina station, liked to keep his hamburgers warm by placing them in a 'cart machine'. There is a pinch roller inside that turns the tape at the proper speed. Our radio station was infested with rats and evidently one of them climbed inside for a tasty snack, thereby getting a piece of the meat entwined in the mechanism.

At six a.m, I hit my 'BARRY MICHAELS' jingle, which went something like—"GET UP AND GET GOING"—a musical bed at that point for me to say something like "GET YOUR CRACK OUTTA THE SACK!"—followed by the jingle singers and our call letters. What the listeners heard was... the jingle..and me saying.."'GOOD MORNING!" (Wham! the meat hits!) "IT'S... ....WELL F#@$ ME!!" ...ending, of course, with the jingle singers right on cue! GOD!

I still don't know how I got away with that one. I did get a call about it, just one, and the guy says, "I know you didn't say what I thought you did!" Thankfully, he had been out drinking the night before. Can I get a big amen for Mr. Jack Daniels??

Currently employed in mornings at WYYD Roanoke/Lynchburg Virginia—Internet and BBC Radio Tees voice guy.

Hope this is what you're looking for, Robin.
Your pal,
Barry

Rick Minerd

In 1971 I was a 19 year-old DJ working at what is now WLVQ- (Q-FM-96). We were the FM arm of WTVN-610-AM, then owned by the Taft Broadcasting Company. This entry was about my first day in radio, but I would move on to several stations in the Columbus area and eventually leave the business for good in 1993 to concentrate on a law enforcement career that began in 1986 (I was a full-time deputy sheriff from '86 to '93 and a part-time DJ and morning traffic reporter).

But prior to 1986 I had spent 15 years working full-time in radio as a DJ and later a talk-show host.

My worst experience came on St. Patrick's Day, 1978 when a fellow announcer was murdered in the hotel where our WMNI studios were located. I was on the air the night he was killed in a hotel room just one floor below the studios. That story, as well as many other radio and cop stories is detailed in my new book, "Deputy in Disguise," which can be reviewed at www.rickminerd.com

Here is the short version of my first traumatic experience in radio.... At WBUK-FM-96 in Columbus, Ohio, I was into my first hour, on the first day of my radio career when the studio door opened and the man in a suit introduced himself as the station GM. He said, "I've heard enough, you suck, get your things and get out, you're fired." A few minutes later I called the PD into the studio and explained what had happened and he told me to relax, "That wasn't our GM, that was our morning man, he fires all of the new guys. ***(RM- it's THIS stuff that made the book worth putting together!)***

Since all of that, I have retired from law enforcement, having left as the Chief of Police. Today I am a writer and author of two books.

Scott Mac

Hi Robin,

My name is Scott Mallory and I'm writing about my experiences in radio too. My book is about being downsized and then making the transition from a morning show host to a stay at home daddy, with crazy radio stories mixed in.

I did mornings in Dayton at WXEG-FM and WGTZ-FM.

How is your book coming along? Mine is taking forever. Right now, I'm rewriting and editing. I'm only half way through and hope to be done by the end of March. Are you going to try and get a literary agent or self-publish? ***(RM-what's a literary agent??)***

I'm in Evansville, IN (a little town best known for being the birth place for the creator of Rosanne, and actor Michael Rosenbaum who played Lex Luthor in Smallville) doing weekends and fill-ins on WSTO and WABX.

My wife runs one of the Target stores and I'm a stay at home daddy. I have an eight year old little boy, and a three year old nightmare... I mean my daughter. She is slowly sucking the life out of me lol. ***(RM-never heard it put that way before... but I GET it, I have 5 of those!)***

Good luck with the book, thanks! Scott

Tony Hayes Mathes

First Job Radio Calls-KCLE, Cleburne, TX.

Worked at a small Country station in Texas. First radio job. GM was quite a character and even though I was all of 18 at the time, he brought me a beer into the studio. Owner was also quite a character and one to not take lightly to the beer. Sure enough, the owner comes through the station front door. I see him through the glass and immediately slam the beer into the album bin. The owner sticks his head in the control room, says "hi" and leaves. Took quite a while to dry out the album sleeves and the smell of stale beer lasted for quite some time.

Worst Story.

Worked for an owner/GM who had quite the temper. I was the ops manager for two stations at my second radio stop in Amarillo, TX for what was then KHBJ-AM and KHBQ-FM.

I Could always tell when the owner was hot about something because his ankles would pop as he came down the hall. The faster the pops, the madder he was. One day he "pops" down the hall at a fast pace so I knew he was hot. My wife happened to be in my office having lunch with me, when in he comes ranting about some little thing that had occurred. I don't even remember what it was all about since it was so long ago. Well my wife, a tall Texas redhead who ALWAYS speaks her mind, looked right at him and said, "why do you have to be such a f...wad."
(RM- I LOVE this woman)

I panicked because I figured that would be the end of that particular gig for me. Turns out it wasn't and I stayed for two to three more years and had a great relationship with the Owner/GM. So did the wife!

Currently, I'm on the beach and looking for the next opportunity.
Thanks Robin.

Tony Hayes (Mathes)

Ken Held

Robin, Baby....How I wish I had the time. I also run the oldest privately-owned insurance brokerage in the country, perhaps the world. I'm CEO and it's been in the family since the 1800s. I specialize in ultra-high risk huge premiums for hurricanes, terrorism, earthquakes, tsunamis, et al, all over the world... yup, you get the picture.

Right now I'm like the proverbial one-armed paperhanger and wandering Jew at the same time. My only therapy is my lady Jennifer, my Ferraris and BROADCASTING... they take my mind off the rest of the crap I deal with every day... lol.

I'm glad your career is on a good path, and I'm always here to schmooze with you about the 'other stuff'.....

Take Care of Yourself! Hugs... Ken

(RM- Ken, you've just done what we, that are on the air, TRY to do, every day... you painted a beautiful picture with words alone! Thank you for that!)

David Gray

My funniest radio story didn't happen to me. There was an announcer reading the news at a little station in Maine while we were on vacation and the story was about trucking fees, which the girl twisted to f****** trees. Long pause on the radio. Never heard her again.

Greg Simpson

I had it drummed into me so many times that you do not bring liquids anywhere near the board. As careful as you are, and we all are, accidents can happen.

If I did bring a liquid into the room, I kept it at least four feet away from the board, as a result. Our station had a coffee dispenser, and it also offered hot chocolate. Mmmmmmm.....

On a six hour shift one Sunday night I bought a hot chocolate, brought it into the studio, placed it on top of the cart machine rack, way at the back,on top of a couple of LPs. I then, somehow, managed to hit the corner of one of the LPs with my elbow and, as if in slow motion, watched the hot chocolate do a double flip over my head in a perfect arc landing, seriously, upside down on the board. The impact popped off the lid and I spent the next two hours, between breaks, dabbing at the board with countless sheets of toilet paper, soaking up everything I could find.

Finally satisfied I went home at midnight knowing that I'd fixed the problem. The next morning, when I arrived at the station for my 9 to 5 gig, I saw the morning man working from Studio B, usually used only for one on one interviews, and blessed with a single turntable. The jock was alternating from disc to cart, and tearing his rather thin hair out by the roots. "Oh no", I realized, this was my fault. I poked my head into the main studio, and the engineer had the board completely apart working with Q-tips and fluid.

Apparently, hot chocolate is not just liquid but, also, a lot of sugar.

Overnight, while taped programming was running, that sugar had crystallized, and by the time the morning man got in, the slide pots were totally useless. Nothing to do but ask the engineer if there was anything I could do to help, while confessing my own involvement. He, literally, grabbed me by the shirt and said if I ever did anything like that again, he would punch my lights out. "Okay," I whimpered.

Frank Baum

Would you like a copy of a work agreement I had in Arizona? Mythical quality. "Can do anything you want with the show as long as the ratings are 7.0 + starting from their actually good 4.7" ??

The new owner/mgr/PD bunch didn't like part of the "anything." They didn't back up the agreement, so I was out with a 10.8 still running. It wasn't counted as paying dues or being nice about it.

"The work agreement on paper, signed, is the mythical equivalent of the unicorn, Robin." Such things have happened. The sad story of the disproving of the reality of the Unicorn is probably not the story you'd like in this book. You'd like people to chuckle happily, if not guffawing spontaneously. Work agreement and "anything" + guarantee of a job = 'wow factor'. "The rest ... not funny." (***RM-one of the smartest men in radio, folks.)***

Jay Davenport

1. My first radio Gig was working as an intern for WBAM in Montgomery, Alabama.

2. My radio horror story goes back to a time when i was a young, 18 year old intern working for a top 40 station (WBAM). One cold January eve I get assigned to set up our Friday club remote, which basically meant I would drive the van out to the club and babysit the DJ while he/she got drunk before the end of the night. Around 2:45 in the morning as I am loading up the van, I close all doors and lock everything up, then head inside to check on my on air talent and get my pay from the nightclub manager (or at least a cut of what the DJ made). On my way back to the van, I went to grab the keys, only to notice they were not in my pocket, but swinging from the rearview mirror...(DOH!!!)....

I call the PD, who chews me out for a good 20 minutes, then instructs me to find a 24 hr locksmith, return the van to the station and turn in any keys to the other vans if I had them as I would be let go on Monday morning.

As I finally pull into the station parking lot at 7 am (it took 3 hours to get a locksmith) the PD rushes out to greet me and ask where the DJ that rode with me is, my job was saved the moment a drunken night jock emerged from the van covered in a hangover induced vomit and his pants around his ankles.

3. I am now morning drive board op for the John Boy and Billy Big show at WXFX in Montgomery, Alabama.

Jamie Gordon

Hi sweety....

1st gig WCTO

WORST EXPERIENCE: Also at WCTO.... So I had prepped all week for my first official "on air" overnight shift at WCTO. I over prepped, but forgot one really important thing—I now know it's called a positioning statement when you first crack the mic. WCTO's statement was Cat Country 96, Continuous Country favorites. I didn't get taught that part and being a total newbie to radio, had no clue. So I open the mic and I am back on stage in 3rd grade for a flute concert where I just freeze-TOTAL STAGE FRIGHT! So nothing comes out of my mouth and I pull it together after some dead air to hit the song...... At that moment I looked across the hallway through the studio window at the night guy and yelled, "I fuc$#d that one up real good! I will never forget the terror on his face as he signaled that I left the mic on and the pod was up—live on air first shift ever, and ironic that I am still doing radio.

WWNU New 92-Afternoons in Columbia, SC

I am proud to say I am at a new gig!! mornings at WXCL The Wolf!
(RM-Congrats!!!)

Slick Show

I'm not living in an environment that's conducive to productivity and creativity, so I'll write more later."

I was on "Power 1450"/KKPW in Tucson...I recorded a phone call, made some edits, and then aired it over the outro and intro of a couple of songs. What I forgot to take out was...

Caller #1: "...no shit?"

Me: "None, whatsoever..."

Only one listener called about it...

Caller #2: "S!ick, did you hear what he said?..."
Me: ... "What did he say...?"

Caller #2: "Uh... oh... never mind... have a good day!"

I've always practiced the policy of AYS ("Are you sure?")...and: if you don't make a "big deal" out of it, maybe no one else will.

Michele Amabile Angermiller

My first job was board op-ing a country station-STAR COUNTRY, on WOBM-AM. My first F/T radio job was nights at WHTG FM, 106.3, a heritage alternative rock station in 1992..

When I used to work at WHTG FM in Eatontown, the board was held together by tape and paper clips. I kid you not. We had two cart machines that always stuck, and sometimes a one minute commercial was on a three minute cart. You would have a 5 minute stopset, two cart machines. You hit the first spot-and then the second...... with your finger. Well, what a nightmare when the second spot was over and the first cart was still REWINDING. Sometimes you would hear the "whirrrr" on the air. Oh, and then there were the owners two cats in the studio!

Today, I am a swing shift on air talent at 107.1 The Breeze, Neptune, Press Communications, and I freelance for The Trenton Times..

On air name: Michele Amabile

Marlene Blair

My first radio job, I opened the mic and did a bit which I flubbed. I turned the mic down (or so I thought) and said "I sure f**ked that up".....Imagine my horror when I saw I hadn't turned the mic off, just down enough so only a few people heard (I HOPE)!

Another one from another station that was still playing records (early 80's) I put "wildfire" on and ran to the bathroom. When I came back to the studio it was stuck on "She's coming....She's coming....SHe's coming...".

Turned down the volume and said..."Well I'm glad she's having a good time but we need to move on".... LOL

....and remember....You haven't worked in radio until you've been fired at least once!

Bob Domingo

has had 2 nightmares within the past 4 days where I'm back IN radio at my previous station in Erie , PA. One involved taking back my PD job only to find that the guy who they had hired when I was laid off, hadn't reconciled the music logs in over a year, and the stack of logs was huge. The second dream was similar, in... which I woke up literally sweating out of fear that I was forced to go back there to work....(insert sigh of relief)

Tom Flammia

It was 5:30am and I had just started my show. The coffee had yet to kick in. A big fat fly was buzzing around the control room, and landed on the console... so I flicked at it with my finger... perfect shot, the fly was knocked silly and landed on the turntable that was playing John Denver's "Calypso"- I thought "no ...way is that fly going to get impaled by the needle- sure enough- the little bugger got impaled and let out a buzz that sounded like a tremendous fart... right over the airwaves. The tone arm flew across the record, and what did I do? I opened the mic, said "Excuse me".... and played a couple of PSA's until I could stop laughing and cue up another tune. To this day I can't hear that song without laughing!

Vince Santarelli

Okay here goes. First radio gig was at WCRV in Washington, NJ, 1580 on the dial. I was all of 17 years old. One day my father said to me, "look, if you really want to get into radio, why not go over to the local radio station and try to get in." So I did, and son of a gun, I became a gopher at the station. My very first day on the air was April 3, 1971. I did the “God squad” from 6 AM to 10:50 AM and did my first deejay show from 10:50 to 11 AM. God was I nervous. But I actually became a star at my high school. Even had a picture of me on the air at WCRV in my senior class year book.

Worst on air experience also came at WCRV. This was sometime later and I had been promoted to the mid day guy at the station. From 10 to 11 AM every day, the General Manager, Nick DeRienzo, God rest his soul, did a talk show and took calls from the listeners. It was my job to answer the phones and let Nick know which line was for him. Well, the phone rang, I answered it and the person asked to talk with Nick DeRienzo.

Well, I held up one finger for Nick to let him know that line one was for him. When he answered the phone, the person on the other end said, "Hello, Mr. Derienzo, this is the Gates Transmitter Company calling about your overdue payment. Needless to say, I wanted to dig a hole and crawl into it. Fortunately, he got out of the call and never, never, never said a word to me about it. I guess he was as embarrassed as I was.

Where I am today. Well, way back in 1985 I got out of radio. Figured I hadn't made it to New York by then, I probably never will. Went to work for the United States Post Office. After 23 years there, I retired and went to work for a bus company driving school busses. Good job, and I'm happy. Oh, and I still do my online column about New York radio, Apple Bites. ***(RM- "gotta' LOVE it! Thank you Vince, for the comment on the NY Radio message board, on my behalf! http://www.musicradio77.com/wwwboard/messages/371688.html ... too bad he didn't listen!")***

David Sherman

Hi, Robin!

My first radio gig was at WMMM-AM in Westport, CT.

I have two quick stories, both nightmares with "small n's," one is more of a "non-nightmare."

If I may...

On one occasion, while at WVIP-FM in Mount Kisco, I got a nasty case of the hiccups-during my airshift! I couldn't say a thing without hiccuping! If I had a tape rolling, I might have considered sending it to Dick Clark, who was doing his Bloopers TV show at that time.

A few years later, I was at WGLI-AM in Long Island. This was the very last radio station I ever worked for that still used vinyl full-time. The turntables were positioned on either side of me. On this one occasion, I had the misfortune of aiming an especially vicious (but, thankfully, equally "dry") sneeze in the direction of the "live" turntable. I watched in absolute horror as the tone arm quickly rose off the spinning record, only to ever-so-gently drop back onto the very next groove. Sure; there were about 2-5 seconds of dead air. But when the tone arm finally landed back on to the record, it didn't miss a beat!! (And take me at my word, when I say "you can't make this stuff up!!!")

(I'm among those displaced by budget cuts; after three years, I'm still looking for a new radio gig. My last full-time "port-of-call" was an eight-year stint at WMNI-AM in Columbus, Ohio.) I'm currently in between radio gigs (casualty of budget cuts) and continuing my quest for a new challenge in the business...

Thanx!
DAVE SHERMAN

Michael Steele

Everyone has it. Trying to get to the studio and your legs don't work. The song is ending and you can't get to the door. I haven't been on the air day to day in almost 12 years and I still have the Dead Air Dream every 6 months or so.

Joe Dawson

Robin..great idea...

I was putting on Mufflers at United Muffler Shop in Omaha Nebraska on a cold Saturday afternoon in October of 1975. The owner of the shop used to play the Big Joe Polka Show on the radio in the shop, and I would always give him hell for it. As I slid out from under a Buick with a torch in my hand, filthy I looked up at him and said.. "I give your polka guy is smarter than me because he at least gets to play that crap from a warm studio and I am here laying in the snow cutting mufflers off cars" ...He said.. "You're right Dawson, you could NEVER do what he does because it takes skill and education that you don't have"...I told him I could be a dj if I wanted to, and he told me to put my money where my mouth is and bet me 100 dollars that I COULDN'T find a radio station to hire me as a dj before the end of the year. I took the bet.

I was 19 years old, and had no college to speak of, and had never thought of being a "dj"..but I don't like to loose, so I went home and wrote down every AM station I could hear from my Mom's kitchen table radio around Omaha. I knocked on doors, made calls and drove to stations all over the area until I wandered into radio station KOTD, high atop broadcast hill (as they put it) in Plattsmouth, Nebraska. Jo Warga was the owner and the entire station including transmitter was at the tower site in a building about as big as a garage. Jo asked me to come into his office where I told him I wanted a job as an announcer. He asked me for an audition tape, I told him I didn't have one with me (I didn't even know what one was), he said, "Fine," and gave me some UPI copy to read to him...I did, and he said "you have a fine speaking voice, do you know how to run the equipment?" I said, "sure." (I had never seen a control board up close)... He then asked if I could work on Thanksgiving morning...I said "YES," and he agreed to give me a try.

The station was a day-timer, so I came in the night before and took crib notes from the announcer that was on the air and got ready to come in and sign the station on low power at 6 am the following morning.

I got it on...was horrible early on, but got the hang of it by 10 and Jo Warga was waiting for me after I got off my "shift" at noon...he said that it was a little rough early, but since it was a holiday it was a good time to break in...he said I sounded fine and offered me a Noon to 6pm job...ripping and reading news and farm markets from Noon until 3 then running a dj music show from 3 until 6..6 days a week for 535 dollars a month.

I took it, won the bet and stayed there for 3 months until I was hired by a top 40 station in Sioux Falls and market by market until I got to Chicago and a long run at CHR B96 in Chicago for the majority of the 80's. Then on to programming until radio became a job of programming computers...when I left the business.

OK...I'll admit it...when I was working in Rapid City early in my career, I used to call the weekend jocks...yes plural..on the competition and tell them "I was "Bob" the new engineer at the transmitter site, and I needed them to hit 9 lower immediately...well, if you are as long in the tooth in the radio biz as me, on the only remote transmitter panels this almost always turned the transmitter off...and it did...and stayed that way for at least 20 mins until their PD must have called to get them to turn it back on...
(RM-you are SO BAD!!!)

I did this every weekend for 3 weekends in a row until they finally must have told the weekend air staff to look out for "BOB" from the transmitter site calling on their REQUEST LINE...yes, I know...I was a naughty dj for doing that...and I am sure the statue of limitations has run out...whew!

I now own a sports marketing and advertising company in Atlanta, Georgia!

Steve Clem

I like the Make-A-Wish part! I like to write.

For the record, I don't think I'm smart enough to be Frank Baum's friend. (Not that he would accept me as a friend, lol) ***(RM- yes you are, he welcomes ALL, in spite of our weaknesses!)***

When I wrote about mine for R&R, I did it as an hour- by-hour diary of that day...beginning with the 8am Dept Head Meeting where no one was particularly interested in anything I had to say....the call from the Ops Manager telling me he wanted to meet with my staff individually that afternoon in 15 minute increments (which I then arranged). My thoughts during the loooooong walk to the parking garage with my box of stuff. And, ultimately, me back at home, sitting in my recliner JUST in time to hear the format change...when my station went from Celine Dion to Nickleback...with juvenile-sounding rock imaging that made the whole package sound like a mid-July April Fool's joke.

My first day on the air at a small AM station in Duncan, Oklahoma, KRHD,

(The listeners called it "crud") was the fulfillment of a lifelong dream. Granted, at 19, maybe it just felt like a lifetime that I had been practicing my patter on our family's Silvertone radio/record player console in the front room.

I was keenly aware that the dream was beginning to materialize as I began my first solo air shift. The format was Country, which I knew absolutely nothing about. But, this was how you paid your dues...how you worked your way up to the big time (and a format you liked) at KAKC in Tulsa, or dare I dream BIG? At WLS in Chicago.

That first day, I was just trying to keep something on the air, and, as it turned out, trying to keep from passing out in those seconds-that-seem-an-eternity... between turning on the microphone and actually speaking on the radio. But, the song finally ended, I got through it and went into the commercial break.

Just then, the door to the control room flew open, startling me (it wouldn't take much on this particular day), and a tall grey-haired gentleman burst in and announced in his gruff Oklahoma-Texas border drawl, "Hi, I'm Lester Agee, the General Manager." Having been hired by the Program Director, I had never had the pleasure, so I squeaked out a sheepish "I'm Steve," while trying to make sure I heard the cue to punch the next commercial on the three-stack cart deck. He got to the point. "Let me see that last record you played," he barked. "Sure, Mr. Agee," I stammered as I slipped the 45 rpm single from its sleeve and handed it up to him from my seated position at the console. He stood there, holding the vinyl record in front of him at just-the-right bifocal distance to study the label. Then, in one sudden motion, he SNAPPED the record over his knee, breaking it into two pieces, and handed it back to me. As I sat there, stupefied, he stated, "We don't need this kind of communist propaganda on my radio station." He turned and walked out.

Now, another 19-year-old aspiring dj might have taken this first day incident as an omen. Or, have been affected in SOME negative way. But not me. I was living the dream! Plus, I had to pronounce "Pottawatomie County" in a weather forecast. ***(RM- Great story! Painted Perfectly!)***

I'm now Operations Director for Public Radio Tulsa and President of Perfect Mix Music Logs LLC!

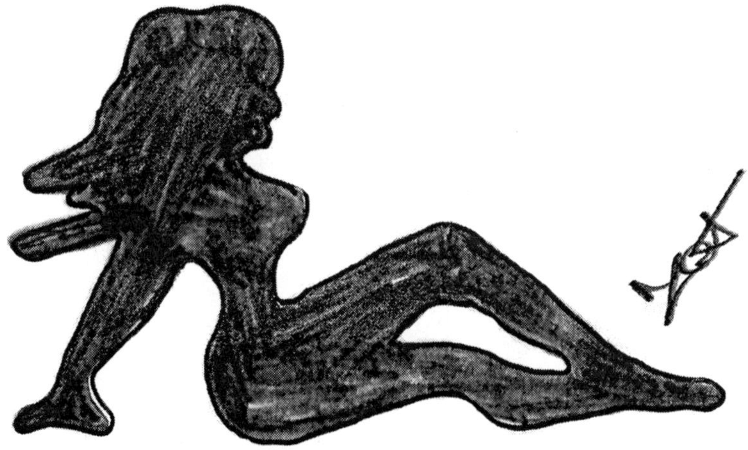

Charles Murray

First worked at WKNC in college ,NCSU's student station. Following that I worked at WPTF/WQDR from 1984-1987. Next was WZZU 93.9 also in Raleigh,1987-1989 as well as Triangle Traffic Patrol Broadcasting during the same time frame.

"My most embarrassing moment was at AM680 WPTF. I had a friend on the phone and was chatting with him while I was on-air. It came time to put in the weather cart, put it in the rack & turned up the volume. Then, "OH NO"... it wasn't the weather, but I had in fact recorded my friend & I talking on the phone about dating this hot girl I was chasing & where we'd be going out to party that night. These were in the day when once you started one of those "carts" you couldn't stop it until it finished. I even went out into the hallway & tried to stop the master control tape & edit it out to no avail, LOL!

Later down the road I told my bosses & we all got a big laugh out of it." ***(RM-did you ever get the girl?)***

I'm currently out of the industry I now own & operate my own Landscape Company & ATV trail park.

Rudy Von der Heide

Under the Radar …

I wanted to get into Radio because I believed in freedom, truth and the 1st amendment. I also thought the sound of my voice could be lucrative and I could live off it. (Insert raucous laughter)

I pursued radio from the most direct route, Community College. I got my degree in Radio Production and started working for 1310 AM WVIP, an AM news station owned by the intrepid Martin Stone. Martin was 100000 years old when I met him. He fired me one morning because the station had failed to air at 6am because the transmitter had died. It was in a locked shed about 100 yards from the radio station under 4 feet of un-shoveled snow itself. 12 minutes later upon realizing I was not responsible he rehired me on the spot. I spent 12 hour weekend shifts rerunning Pat Buchannan tapes on reel and playing four commercials. I can't remember what they were for. I ran tapes there for about 4 months until I found another job driving limousines. A few years later the station in Mount Kisco burned down, not because of me.

Cut to 6 years later I found myself working at 102.7 WNEW for Radio Chick, first as a phone screener parlaying that into a floating full time producer position. I had the best job. Hang with guests, play foosball with Anthony of the Opie and Anthony Show during commercial breaks, (Opie and Anthony were the princes of afternoon laughter at WNEW NY and building a presence in national syndication) make promos for the show by isolating a few 20 second clips of daily hilarity for station promos and that was it.

The worst Radio Experience was 9/11/01. The morning attacks played out for me on "Sports Guys" (among other media sources) and I wondered what happened to the EAS? I never heard the tones. I stayed home that day for the first time in almost two years of work fearful that the towers were just the beginning. The day after three hysterical French Canadian Women had cried their way into the green room saying they had no money and were trying to deal with the fact that their vacation plans had been ruined by the destruction of the towers. I remember being asked for a cigarette by one of these hysterical Canadian chicks thinking "Jesus! 3000 people just died and this woman is conning us for free Amtrak tickets to Montreal!"

As days passed open phone lines for the next few weeks was all the talk hosts could do. Two weeks listening to the voices of the country all in relative mourning and angry with confusion. Answering phones, listening to the calls of people still searching or hoping loved ones could still be recovered or who was to blame for it all; probably the worst thing that happened to all of us.

Considering what I'd witnessed hiding in production studios and under stairwells in the many Radio stations of CBS for the last 10 years I wish I could say the dream has been realized. I stay under the radar taking soulless board operator shifts considering the industry appears to be surrendering to the internet's interactive entertainment, news and social networking.

The best tool for the first amendment should still remain unchecked by an ever reaching federal mindset and I hope one day the miracle of Radio will save us from ourselves. ***(RM-Rudy, who followed me every Sunday night, reminds me that we should not take ourselves as seriously as we THINK we should!)***

Crista Leigh

I was 19, at my first on-air gig at WCDB at UAlbany. My air shift was 2am-4am and I figured that next to no one was listening. Before I "found my personality," I tended to yell quite a bit into the mic, or speak low and (what i thought was) sexy, really close to the mic. One night, I got a little too into it and bit... down on the mic—ZAP! I screamed and campus security came running into the studio. Never heard the end of it.

Believe it or not, this has happened to me at least a dozen times since... somehow biting down on the mic.. ouch! ***(RM- my son would say, 'that's what SHE said,' right about here?)***

Mark Vaness

My first full time FM radio gig at KSFM in Sacramento, Ca. I did over nights, helped produce the morning show, worked part time in the engineering department building studios and making repairs while helping the promotions department, keeping vehicles clean and remote setup. On top of doing my own appearances, I was tired a lot! The morning guy was living inside "Santas Workshop" on location to collect toys for kids. I was cleaning up and taking down the equipment. I hopped into the Coca Cola Video van and pulled away from the site with our Big Red Radio Boom Box studio attached and forgot to retract the Marti antennae...it caught the power wires to the Workshop (which was about 15 feet off the ground) and pulled it over...luckily the cable broke and it settled back into place! I almost killed our morning show host!

That was a bad day... 23 years later I am the morning show host at (WBZZ) Buzz 105.7 (Hot A/C) in Albany-Schenectady-Troy, New York.

Charlie Morris

Everybody's had the dream: Can't find the record...dead air...the track won't play. However, here's the amazing part; not only do we all share this dream, but why is it so persistent? I have been off-air now for over ten years. And as for the dream? It still occurs at least a few times a year, and unfortunately it occurs much more often than that Pam Anderson dream. Damn it!

Karen Lee DeFreiz

Hey Robin –

My first radio gig was at WSHC, Shepherd College, Shepherdstown, WVA, back in 1984, playing rock. My first paying gig was 7-midnight at WYII in Williamsport, MD, a country station. That was 1986.

I was doing swing for our afternoon guy one day at WJXA in Nashville when a really nasty storm blew in. We had a tower just outside our studio window, literally just a couple of feet away. I had the mike open, was on the air, and heard this loud sizzle in my headphones. I shrieked, threw the phones off my head and watched as the whole board started smoking. Lightening hit the tower and fried the entire studio. Smelled to high heaven! I thought my hair was on fire.

Back in the day, (late 80's) at one of my previous stations, WAYZ in Waynesboro PA, we read the news live behind the board every afternoon. After I got off my midday shift, I'd put the stories together and deliver the newscast for PM drive. So I'm standing behind the board, reading it live when one of our part time guys sneaks in with a lighter and thinks it would be really funny to light it on fire as I'm reading it, which he does. Speediest newscast in my entire career!

I currently work for Premiere Radio Networks - Nashville as an Affiliate Relations Director. I've been in this position 11 years. I started with MJI Broadcasting in 1999 and we were bought by Premiere a year or so later. Previous to that, I spent 15 years in radio and 4 in indie record promotion. Good luck with the book Robin! Let me know when it's out! If I think of any more stories I'll send them over to you. There was that time I had to jello wrestle for a promotion..... ***(RM-amazing the jello didn't catch on fire!)***

Michael Flynn

Worst on air fopah...First job in '86. News anchor opposite the PD. I'll tell you what I thought I said... "His opponent says...Roberts is giving his constituents the biggest SNOW job since the blizzard of 1888!!!

As I noticed the PD on the floor laughing...I thought...what the hell was so funny???

One of the funniest parts for him was...He knew I had no idea!!!!!...He KNEW I had no idea!!!

Lana Wilder

My first radio gig was back in Oct 1985, at KLKT / Incline Village, Nevada (on the east shore of Lake Tahoe.) I had lived there for 5 years prior (ski bum!) and hooked up with a guy there who did mornings at the station. I followed him when got a morning gig at WGFM in Schenectady. The Capitol Region of NY! I thought I was moving to a wonderful metropolitan area, full of excitement! NOT! The silver lining was that 2 years later I ended up going to a radio school there where he taught, and after 10 weeks of learning what buttons-did-what I ended up heading across country again and back to Tahoe for an overnight gig at his old station. It was the perfect first radio job. A full-service AC that played everything! And I do mean everything. Oldies, 60's, 70's, the currents of the day, and there was even a smooth jazz position in the clock. ABC network news every hour on the hour, so I had to back-time into it. A friend and former co-worker of my old boyfriend helped me get the job, and let's just say he "exaggerated" a bit when it came to the amount of experience I had. I was so terrified my first night on the air I BEGGED the 7-midnight guy not to leave me alone.

This was back in the days of records and carts, and I remember that we had two triple-stack cart machines and one of the slots wouldn't hold the cart in it—it would actually fall out and stop playing unless you taped the cart in with a piece of duct tape.

I was dating a very handsome older man who happened to work the same hours as I, and he would stop by so we could make out during the songs. This was great except for when the building security guy would walk by and catch us.

My PD was a bit of a drinker (!) so he was usually up all night listening, and often I'd get the dreaded hotline call if I goofed something up. Lake Tahoe is known for some very heavy snowfall, and if the satellite dish on top of the building got too much snow build-up, we'd have to push a button to raise the dish to dump the snow, and if it didn't come out we were supposed to climb on the roof and use a broom. I refused to climb up on the roof. I don't do roofs! ***(RM- can't say that we'd blame you here Lana!)***

I only worked there for 6 months before heading back east again for my next gig, a CHR midday job. But I have such fond memories of that crazy place!

Oh yeah, I remember the morning guy at my first-ever radio gig was doing a live read and was supposed to say "polyester knit shirts", but it came out "Polyester nerd shits".

AND...

It was the early 90's, I was on my 2nd tour of duty at 95 Triple X and I was doing a Saturday remote at one of our malls in Burlington, Vermont. Back in those days we used to blow up at least 500 helium balloons during a 3-hour appearance, every freakin' kid wanted one, and it was hard to get away to do anything...on-air breaks, bathroom breaks, etc.

Fortunately I did have an intern there to help me (Sweet young 20 year -old girl) and all she could talk about was how much she was craving a Frosty from the Wendy's that was in the mall. Finally, 2 hours in, there was a slow-down in the balloon action so I told her to run to Wendy's and get her Frosty. About ten minutes later she comes back, sooooo happy to have her beloved treat, when a family with a fat kid walks by and the kid threw up a Frosty right in front of her. End of Frosty craving for the intern.

Today, sadly, I recently (2 weeks ago) was let go after 24 years of employment in Burlington/Plattsburgh/Montreal market, 7 on the same morning show. Long story, bottom line: I made too much money and they found someone younger who would work for half of what I made. Vermont is a "right to work" state so an employer can get away with all kinds of nonsense.

Not sure what comes next, I have had interest from other markets and found out I have some fans in high programming places. I know whatever comes next will be better than my last gig, which, aside from the morning show, sucked!

Larry Yurdin

It was back in the 60's. I'd just been hired as program director of a station that was converting in stages from easy listening (elevator music) to free form rock (at that point daytime elevator, nighttime rock). I was hired over the head of a nutty character who knew nothing about radio but had conned the owner into hiring him as a consultant (remember, these were the 60's). This guy was determined to get rid of me so he dosed me with LSD without my knowledge and dropped me at the station on a Saturday morning saying things like "This is your station. Do whatever you want to do". I went into the control room and switched to rock. Between cuts I'd turn on the mic and say things like "Don't Worry. Everything will be alright.". Needless to say I was fired, but eventually the truth came out, the "consultant" was fired, and I was offered my old job back but by then I had moved on to other projects. Instead I maintained a relationship and returned to town to do occasional air shifts. ***(RM-I think we'd ALL love to know who the Consultant was!)***

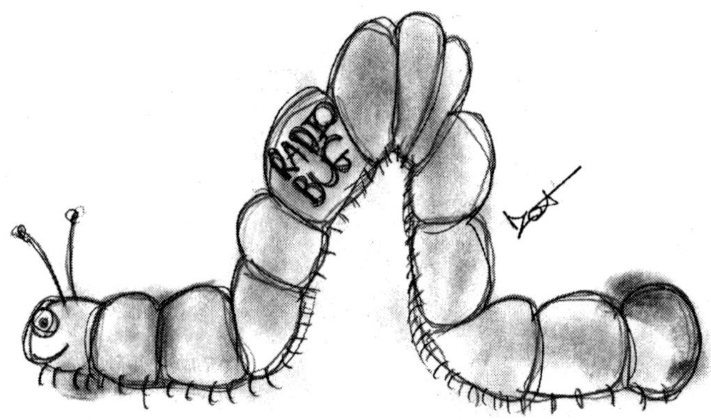

Phil Lacefield Jr.

My First Radio Gig:

In 1983, I had just started high school in Middletown, Ohio, and become active in Junior Achievement as an after-school activity. One of the "corporations" available was a 30-minute weekly student-run radio show on WPFB-AM, to be produced, marketed, sold and operated entirely by JA students. It sounded like fun, so I signed up – and caught the radio bug right away.

We were only required to be at the studio for an hour or two a week, producing and airing the show live, but I managed to find myself there every day after school (the studio was on property right next to the high school), learning the ins and outs from the pros who seemed to 'take a liking' to the kid asking all the questions.

Eventually (as these things usually go), I was in the right place at the right time and was offered the lowest of low-end gigs, a stand-by weekender. With my parent's permission of course, as I was only 14 at the time, I managed to work several late-night gigs on both the AM and FM side of the station; helped out at remotes and high school football game broadcasts; poke about inside the guts of several transmitters under the watchful eye of a cranky but loveable chief engineer; and discover that radio sales was just not something I ever wanted to do for a living.

The Story:

At WWSN (now WMMX) in Dayton back in 1990 and 1991, we did a spectacular year-long giveaway of a 4-door '57 Chevy Bel Air. As the assistant engineer/remotes guy/fill-in grunt/car guy at the station, the care and feeding of this automotive monstrosity fell on my shoulders. That car spent every minute of nine or ten months with me, and was basically my primary transportation - our GM at the time thought it would be good to put out the ballot box and stanchions everywhere it was parked, so every damn time I had to go to the store or the post office or whatever I'd have to drive that car and put those things out. It was fun for the first four months or so, but after that grew tedious quickly.

Thing was, the car was a total piece of crap. It had been purchased sight unseen by a stereo shop in Dayton (the primary sponsor of the contest), and they'd put what was at the time the highest-end stereo system possible in the trunk, so that the tunes certainly sounded good when you were driving it. The problem was, no one bothered to check and see how structurally sound the car was - turns out it was all ate up with rust, all up and down the body. I quickly discovered that someone had given it a quick respray to cover the crappy bodywork, but I was always in fear of the frame snapping in two from the rust – to say nothing of the mechanical condition of the engine. And the transmission. And the brakes. I ended up spending a bunch of my own money basically rebuilding the mechanical bits, because I got so tired of it breaking down on me all the time.

Finally the night came for the drawing, held the Saturday night of the Labor Day telethon. We'd been collecting ballots for almost a year at that point, and had several hundred thousand. When the name was drawn, I was producing our weekly Saturday night live show, and was listening in to the televised feed of the live drawing that was playing on the bar's big screen TVs. I could hardly contain my glee, knowing this abomination would soon be removed forever from my life, never to be seen again. Finally, Dan Edwards (our midday guy) reached into the gigantic drum of ballots...and drew out the single entry of my aunt's husband. I made a mad dash to the phone to tell him he'd won, but also to remind him that if they asked him if he was a relative of a station employee, he didn't know me. Only later did I realize that I kinda wish he HAD told them we were family, because he kept that thing until the day he died, meaning I had to see it every time there was a family gathering. Fortunately, Don was quite the car guy, and rebuilt the beast from the ground up, making it an impressive and well-known classic amongst car show attendees. After his death in 2001, my cousin kept the car in pristine shape, and only sold it outside the family in 2008. ***(RM-Phil, you need to adopt MY philosophy about cars, you'd probably be much happier, 'Don't care WHAT it looks like just put gas in it, and gimme' the keys!")***

Dan Wing

While I have many screw-ups that are probably worse than this one in my 4-year radio career, the one slip that keeps coming to mind happened while working play-by-play of a HS basketball game 2-years ago. The quote is one of my favorite's- "Samuels steps back up to the free throw line for his 2nd attempt at the charity stripe.......he missed number 1, but if he hits shit number 2, he'll put Egan(his team) up by a deuce." I realized what I had said right away, but as we all know, in most formats you just blow past screw-ups and hope nobody noticed. I couldn't wipe the smile off my face till half time, and have to admit that I was actually pretty proud of the fact that I had made such a high-quality mistake.

Mike Marchinuke

WNCO FM, Ashland, Ohio, 78 or'79. On the phone with the first of my ex wives. In an argument. Static turns on the mics and I don't know it. Mouth and phone close to mic. Soft song. Used words that sounded like bunt, witch, sock, and I believe truck. Multiple times. GM walking past speakers in lobby looking like "I'm not really hearing that". Chevy Chase outro and moved on. Somehow kept the job.

Sandy Sloan Collins

I had a "stalker" too. Sweet guy in Dayton, OH. 7-12mid, lots of time to chat... no internet, cable or cells! He sent me flowers, cards & music, but he was off. Wanted to meet me. I declined, "company policy." One night, he started making strange sounds. Having a seizure. I yelled into phone for 10 minutes trying to get him, then hung up & called local police. They knew him immediately, told me this always happens when he goes off meds and somebody will find him eventually. Yikes.

Called them back next day to find out if he was dangerous. Nope. Just crazy,ma'am. Once found him at the airport, trying to take off the runway... without an airplane.

I didn't hear from him for a couple of months, and then, BAM, he shows up in person at our 4th of July event. 20,000 people and 1 whack job. Scared the bejesus out of me. Glad he wasn't dangerous. I just told him that I wasn't interested,and he went away. Just that easy. Poor guy. Wonder where he is now. Maybe runway 2L. ***(RM Hahaha!)***

Jaye Albright

Really. My radio nightmare is the same we all share, truthfully and that just didn't seem book-worthy. ***(RM- the fact that you admit to HAVING it, like the rest of us, is enough!)***

1. WSOM-FM, Salem, Ohio was put on the air in the late 1950's by the local TV repair guy, Russ Jones and I started hanging around the station from the very start, loving just being close to radio. Ted Nearhouse, who later launched WDBN-FM, Barbertton-Cleveland with great success was the first PD and kind enough to look the other way. Night guy Chuck Krouse (who became a legendary news anchor in Kane, PA, until his retirement) did a trade deal with me, teaching me the equipment in exchange for babysitting services until my mother blew the whistle on that, embarrassing both of us. ***(RM-today we'd call that 'entrepreneurial!)***

When I entered Kent State University in 1961, I got hired by GM Mickey Neikirk at the brand new WKNT-FM. It ironically broadcast big ten inch reels of music from "IGM" in Bellingham, WA., which all came full circle in the late 1980's when I went to work for Les Smith's Broadcast Programming in Seattle, its descendant company.

2. Getting fired by Mickey at WKNT-FM, either because I didn't own a car in those days and thus hitch-hiked to work each day from the campus

and my arrivals at the station were subject to getting a ride OR because I sounded simply terrible. Probably a bit of both.

3. I am consulting partner along with business partner Michael O'Malley in Albright & O'Malley Consulting, lucky enough to work for many of the very best broadcasters in country radio.

See how lame that is? ;-> Thanks for asking. My advice: don't include that pretty average report in what I know will be a wonderful book!
-J.
(RM- in this case Jaye, you are humble for saying this, and a touch of 'humbleness' in today's day and age travels a great distance. Thank you for sharing.)

Paul Evans

My first gig was at Ball State's College Radio station, WCRD 91.3FM and I was the Program Director. I am hanging out talking to our production guy when we hear James Blunt's "Your Beautiful." Well our Music Director didn't pay close attention to which version was uploaded to our

Automation system because the lyric "I'm F***in High" played. ...Didn't hear the end of that one for a while. Luckily the FCC wasn't listening. I now check every song before I play!

I now work for Backyard Broadcasting in Muncie, Indiana while finishing up my schooling at Ball State. I am a producer/board-op for BYBs ESPN station 1340am, and a DJ on BYBs Indiana flagship WLBC 104.1FM.

Con Sofologis

I went to the New School of Contemporary Radio in 78/79'ish. I was one of the first graduates to hit one of the BIG stations in Albany, right out of school. Not sure if it was talent or luck. Went straight to the mids slot on WFLY (92-FLY). Did a bunch of swing shifts on both FLY, and our AM, WPTR. Survived format changes, ownership changes (Rust Comm, to 5 State Tower, to Albany Broadcasting) Kept a part-time air shift, but went into engineering. Did many years of voice-over, spots, liners (the voice of those obnoxious Stereo Liquidation Sales in the 80's).... Became Chief Engineer of WPTR/WFLY, and kept airshift 1 day a week.

Moved to VA as Chief at WVMX, back to NY to WQBK, and then to Wash, DC to WTEM Sports, and built WBIG Oldies. When BIG 103 went on air, they were still short air staff, so I did weekends on BIG.

When I was doing mids on WFLY in the early 80's, this woman would call every night from the tub. She would explain everything she was doing...... I would say "hold please", do an intro or wx, pick up the phone, and she was still at it. I was waiting for some huge guy to start banging on the back door of the old St Albany studio.

In the early 80's at WFLY (92-FLY THE FM Station.. back then). I was running AT-40 on vinyl, and mixed up the segments. It played 30 minutes per album side, and I went from # 35 to # 10, and back to # 25. I think there was also a quick nap during the "God Squad" tapes...

Still do voice-over work for my wife's business. Me. Got out of radio, and went to Secret Service about 14 years ago. ***(RM- having trouble figuring out how you jumped from RADIO to the Secret Service! I think for the next book, we'd ALL like to know that, if you wouldn't mind. One industry wants you to talk, and the other wants you to shut up!)***

Jim McGregor

I did overnights in a small market CHR in Northern NY. We had an automated AM that I had to babysit. So, picture 3 studios separated by glass- My studio, the automated reel2reel room in front of me, and the AM Studio in front of that. Well, the AM studio had a nice sized picture window looking outside to a main street. (Every so often someone would walk by that window and scare the $%*& outta me. Yeah, even @ 3am.) Well, my landlord, who was a local police officer, would park his patrol car outside the window. Just as I would stand up to read the weather- he'd blast me with his search light- temporarily blinding me so I couldn't finish the weather. Not to mention the "Holy $%*&" that came outta' my mouth. Sorry Dick!

Sean Mattingly

It was 1984, my first year in radio. We had reel-to-reel automation. I finish reading the live weather. Hit "start" on the automation and the reel starts, tape instantly breaks. Reel is spinning tape, out of control, all over the room. I scream "I HATE this f*cking job!". Mic was still on. ***(RM-Omg... we FEEL that one, we REALLY do!)***

AND...

I like to play a trick on my new interns. I make the intern bother the DJs during their show. All I do is give the intern a little hammer and ask them to check the tightness of the wood trim on the 5 studio doorframes. They're supposed to tap all the finishing nails down tighter (starting with whichever studio is on air). Waaaugh! So funny to watch the angry/ confused interactions through the glass at a distance!

Tony Doolin

I hate the recurring dj's nightmare... I've been out of the business for about three years and I still have it! ***(RM-THREE? It'll haunt you for the rest of your life! Do I get an "amen?!")***

Bob Aldrich

One of my first shows on WLNG. A PSA for the "Mattituck Fire Department". Well, who knew that ANY town had a 'tire department'. Hopefully, there's a statute of limitations on fleeting expletives!

James Trefney

WADC/WIBZ...summer 1975. WADC was a 5,000 watt daytimer and WIBZ a 3,000 watt FM running TM Stereo Rock. I was a college steudent at Ohio University Athens Ohio, doing weekend nights in Parkersburg. One Sunday night I signed the AM off at 9pm, I then babysat the FM until midnight. I signed the FM off and drove back to ...Athens.Upon arrival back at my dorm at 1am I dialed across the AM ... a big a** 5,000 watt carrier was on the air! Yikes, drove back to Parkersburg to shut the AM off. It seams at times when you signed off the FM, the AM would kick back on. (I learned to check all meters before leaving the station)

Chris Ryan

My name is Chris Ryan - my real name, not a "radio name", and I started my career as a volunteer DJ from midnight to 2:00am at KSDT-FM at the

University of California San Diego. We played stuff like Be Bop Deluxe and early punk and prog-rock (Steve Hillage, Happy The Man, Gentle Giant, Genesis). Of course, there was no pay involved, but I got enough air time to cut up a demo tape, stuff my VW bug full of camping gear, and head out on the road to look for a job in radio that would pay me some cash.

At 21, I began my drive and pulled into my first stop: Tucson AZ. I dropped off my little 2" reel-to-reel tapes to 3 local stations, one of which had just been voted one of the "Top Ten Best Radio Stations In The U.S." according to Rolling Stone readers. The PD, Jim Ray, was also the morning jock. I met with him and we hit it off. They played amazing music! Like NOTHING I had heard in San Diego. He offered me a part time job and, after giving it some thought, I took it. I lived on food stamps the first few months 'til I got a full-time overnight gig. This was June 1980. The station: the legendary KWFM-FM (92.9).

In 1981, Frank Zappa was touring to support his new double album "You Are What You Is." KWFM-FM played Frank Zappa, but since we had been sold to another company and were moving to a more conservative format, our playlist was limited to "Montana" and "Yellow Snow" and "Cozmic Debris." The safer, more well-known material. Well, Frank had always enjoyed coming to Tucson and had a reputation of coming to KWFM and playing "guest DJ" since we were very open formatically. He had a BIG following in Tucson, largely as a result of radio airtime, which eluded him in most radio markets.

The night he came in, I was doing the 7-Midnight slot. I didn't know this was going to happen until I arrived for my shift. We had ONE mic in the studio, so it wasn't set up to be a traditional interview. It was intended to be just him on the air playing tunes and taking calls. I was just going to run the board (and try NOT to get nervous, as Frank was well known to be tempermental and outspoken).

He started the show by saying that his first order of business was to play his entire new album. All 4 sides. I cued up the first cut on side one and my hot line rings. On the other end of the line was my Program Director (who was listening at home) telling me that I could NOT play the entire album, and to stick with the hits. Frank was sitting right there while I took the call, so he knew what was happening.

That first song faded and I turned the mic back on for Frank. He began to complain on the air about how radio was being controlled by some guy in a suit half-way across the country. He railed about corporate ownership and consultants who take the freedom away from programmers. He insisted we play another track off the new album. I was afraid I'd get fired if I didn't follow the format requested by the PD, so I made a "finger-under-the-throat" gesture to try and get him to stop using our station to slam radio. This was my biggest fear: opening up a microphone and running the board for him to get me fired. What to do! He saw he wasn't going to get his record played, so he said "fine, I'm leaving." His mic was still on, live with no delay. He says "I'm not mad at Chris, he's a good guy, but I'm not going to do this." He left some concert tickets behind for us to give away and he walked out. On his way down the hall to the front door (accompanied by his VERY large body-guard) he saw a poster for his new album on the wall. He took out a sharpie and wrote "Play my goddamn record! This means YOU! Frank Zappa" I still have this poster framed today.

Post script: At the concert the next night, Frank was onstage in front of thousands - many of whom had heard the interview. He explained his experience on KWFM and then told the audience to call the "guy in the suit" at the station who was controlling what DJ's played. I cowered in my seat and felt soooooo small and deflated. I thought my career was over. Fortunately, I wasn't fired and we got through it. But my Program Director took lots of angry calls for the next several days.

I moved from KWFM in 1987 after the station was sold and went Adult Contemporary. I took a job in my hometown, San Diego, at KGMG-FM. I ended my career there in San Diego in 1997 as a result of radio deregulation. Jacor bought our stations, laid of a bunch of good people, and started using tracked voices from other markets on many shifts. Radio was being controlled by regional managers - saving money and ruining a good thing. I miss the days of local radio ownership. True independence! I spent 19 years in radio total - and they were good years.

In 1997, I made the transition to Marketing and Advertising, which is what I'm doing today. I live in Cincinnati and work for Dean Houston, Inc. A marketing agency near downtown.

I still have the poster, and the interview itself has been saved as part of an online archive to KWFM. Some really good air checks and photos from the "good old days" of open format Tucson radio. Check it out!

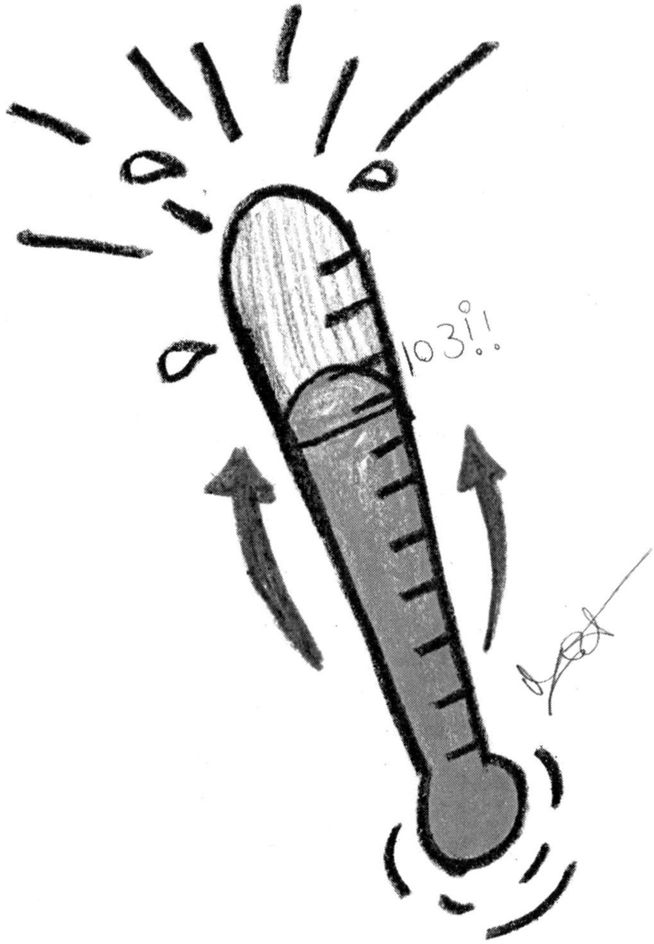

Jim Ray Cobb

(RM- I hope you two are still friends, because it's pretty damned amazing that this story followed THAT story?)

There's a lot that goes along with this (above) story. I was the bad guy. Frank later went on to tell the crowd at the concert to call the radio station and tell them the guy in the suit he sucks or something like that, and they did! The guy in the suit, me (didn't own one) was home with a temp of 100+ getting calls from his GM who was totally freaking out, some BS about the FCC. Funny how S#%# still rolls down hill. More to come about the life and times of K-DUB. What a life it was!

First gig story, 1970 or 71 I walked into a radio station in Concord NC, WEGO ask for a job, the guy said he would hire me if I would cut my hair, it was about as long as yours. Being desperate I agreed. He followed me to the barber shop. After all he was going to pay me $65.00 bucks a week.

Worst nightmare turned out to be the dream job of my life. I got a job at KWFM in Tucson working 7-12. When I got there after driving 2,500 miles from the coast of South Carolina where I lived on an old Plantation, the PD looked at me really funny, "You're Jim Ray?" he said "that's me why?", he didn't say much. Later that night the owner came in and told me I was suppose to be black. First time I ever heard that, freaked me out, thought I was toast. After a few days things settled down and someone told me that since I was from the South and lived on a Plantation they thought I was black. (I guess they forgot about the slaves being freed) Turned out to be the best 9 years of my 20 year radio life.

I've been out of the business for more than 20 years and own a small advertising agency in Charleston, South Carolina. Really miss the old radio days.

Recurring Dream: Record running out and can't get another one out of the album cover."

Daniel Coulombe

"Robin,

I was 19, from a small town in Quebec and it was my first radio Gig in a Northern Ontario mining town... I thought the correct pronunciation was Bruce COCK - Burn and that the City in Arizona was TUCK - sun...

AND...

1980 Midland Ontario. The new kid with the big voice had been working with us at CKMP for about 4 months. One night he suffered a major brain cramp and forgot that the studio audio goes off when you turn the mic on. So now there's dead air... and whatever button our friend pushes on the board, nothing happens. About 3 minutes into this, he starts swearing loudly not realizing that anyone listening can clearly hear him. This continues for a bit until people actually stop by the studio window and attempt to clue him onto the fact that he's live. Listeners are throwing pebbles into the window to get his attention. By this time, he's now very frustrated not only at the equipment but also real pissed at what he's calling 'hicks', assholes and idiots. The station manager is also calling by now, and our announcer answers "WHAT?" ***(RM- Hahahahahaha!)***

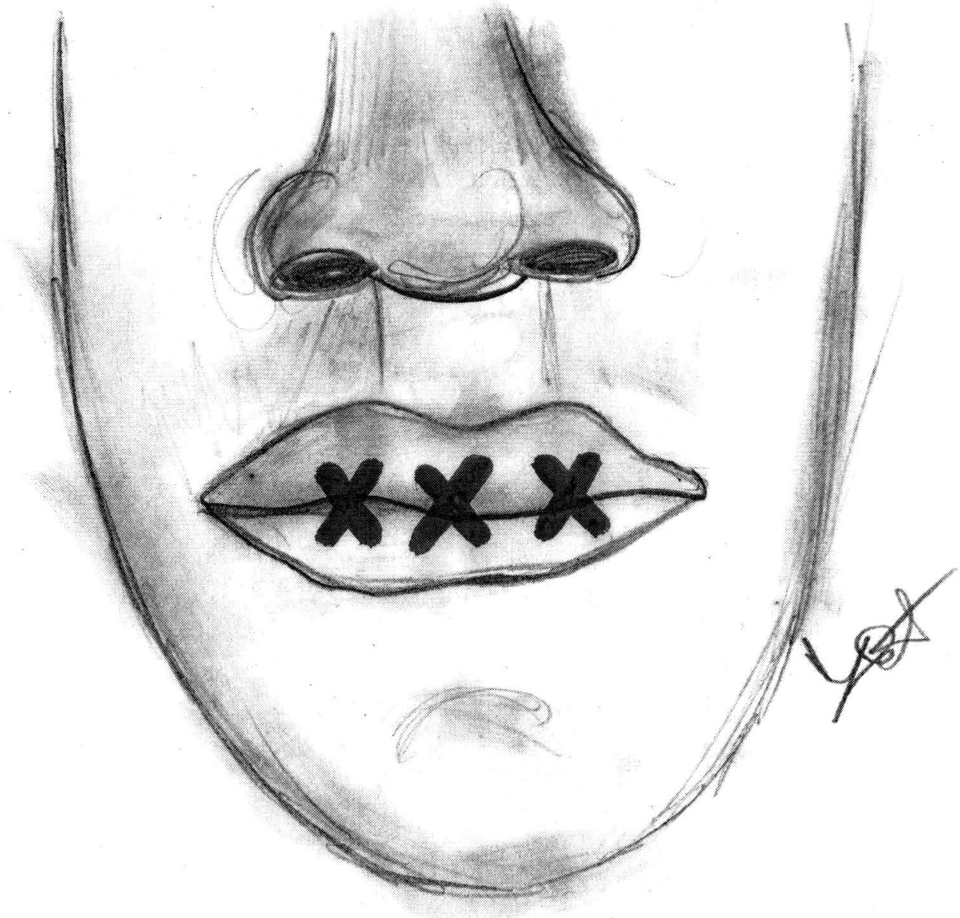

Lou Portanova

Hi Robin.

My first gig was at a small 1KW Daytime/250 watts @ night station. WABY, when it was in Albany, N.Y. on 1400 KHz. The format was an oldies/top 40 hybrid. I started when I was 14 years old. That was in 1973. I got hired to take out the trash, clean the floors, mow the lawn. I spent more time playing in the production studio, than cleaning the station, wanting to be on the air.

Finally got the chance to run the board and remotes. I also ran all of the Sunday morning programs (Religious and ethnic programs-making minor live opens for some of the shows). Those days would start at 6 a.m. and go until 3 p.m. There were some times they would be 7 a.m.-6 p.m. What I would do, is before any of the programs started, I would sign the station on, a little early & do a show.

The PD eventually gave me a chance with a 1-3 p.m. Sunday airshift. -A little side note-When I was running the board for a remote, early one Saturday, there was a live spot scheduled, for an adult movie theater. The lady that was still at the station from the overnight show, told me "just do it", so I did. Yep-here's this 14 year old kid doing a spot for a triple x-rated feature. I think that was actually my first time on the air. I had to put the board in "program" and not listen on the air.

We had a ton of compression and reverb, which threw me off. I was so nervous, too. (ah, the good old days).

I used to go by the name (Brother) Lou Porter. The "Brother" came from actually running the black gospel shows, and also emceeing one of their church functions.

I was at WWOM (Now WKLI), I stacked my spots/carts for the hour. The triple deck was next to the turntable (yes records & carts). I turned fast & knocked the carts onto the record that was playing. (Yes-like the Stern movie, but long before the movie) I potted the turntable pot down, put the tone arm back on the song & potted back up, and then back sold the song as part 1 & 2.

Now to 2010-I do 7 pm-12 midnight at Country 107.7, WGNA-F.M., in Albany. My show is called the Cryin', Lovin' or Leavin' show (Live/local dedications). I've been there for 20 years. It's a great bunch of people to work with and work for.

I now go by the name "Brother" Lou Roberts. Yep-the "BROTHER" has stuck with me through the years. BTW/WGNA has been the number one station for a long time.

I'm married to a lovely lady-Linda. We reside in Colonie, N.Y. (A suburb of Albany), and have been married for just about 3 years. (Thanks Robin)

(RM-No- thank YOU, Lou. Lovely, just lovely. :)

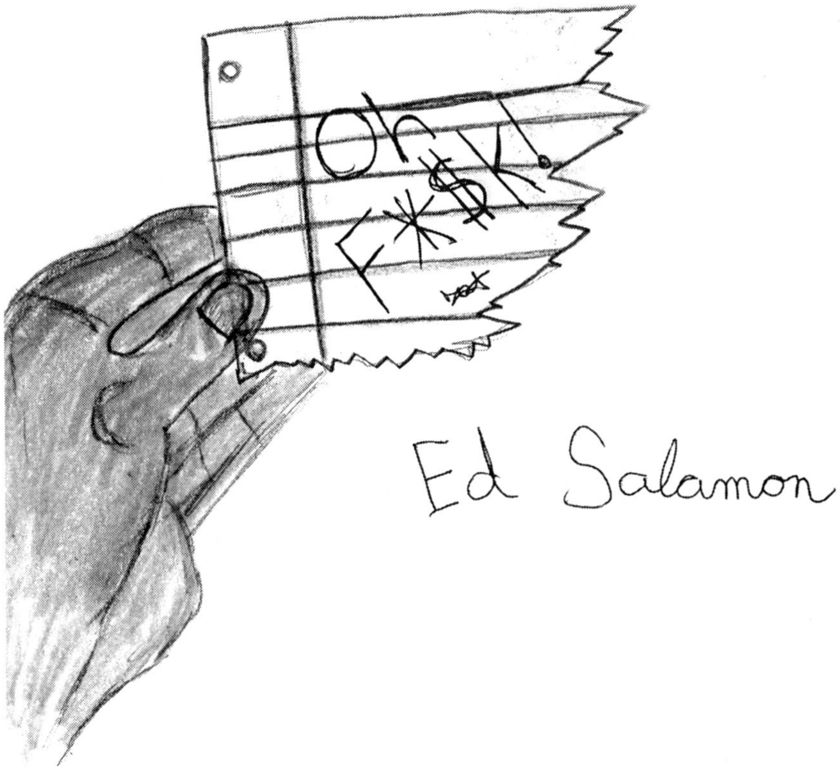

Ed Salamon

When I programmed Country station WHN in New York, I produced a series of live broadcasts from the Lone Star Cafe. One night I learned that even a Country station needs a delay. Delbert McClinton was in the middle of his set, when Elvis Costello joined him on stage. Delbert was so excited, he said "Holy s**t, Elvis Costello is here." Lone Star owner Mort Cooperman told me not to worry, he would handle it and he wrote a note and handed it up to Delbert, who looks at it and then says "Oh f**k, I wasn't supposed to say s**t on the radio", at which point I ended the broadcast. The event was written up in Rolling Stones' Random Notes column the next week, which was the first time I had ever been mentioned in that magazine, and I still have my note from Delbert apologizing for the "bad mouthing." ***(RM-I love you Ed!)***

Russ DiBello

The legendary Gates "Yard" console (so-called, BTW, because it was 36" long... really!) was fed its AC power via a power cord, terminating in... terminals. As in, screw-on, metal, bare naked TERMINALS, right there on the back of that baby.

I learned this factoid, the day I dropped a pencil behind the board, and attempted to retrieve it.

If you sell the movie rights to this to me, please have it written into your contract that I am played by a Top 5 male box office star, known for his sex appeal and wit... oh, you know... George Clooney, Gerard Butler, Brad.. :-) I wanna see Brad Pitt, wearing the Brush Clevite headphones from my first gig..!

I do the kind of radio show where you can't really tell when I'm going off the rails, because that's pretty much where I live, most of the time. What would seem like a horrible error or a situation where the "jock" went bonkers on a nice, safe White People's Radio Station is just another day at the office for me: "Oh, it's him, being silly again!"

Well it wasn't always so. I had to start some where, and where I started, there was a somewhat rigid and conventional Top 40 schedule, or "format" (as we slick pros call it). The guy in-between the songs and jingles and commercials just had to sound professional and "Bigtime"; the latter because the station was in a college town the size of a box of Tide. So, the smaller a station and it's market were, the more that "Bigtime" was a desirable quality.

And this was a formidable problem for me, considering that I: [a] had no experience of which to speak; [b] had no voice that could speak, at least in the bombastic, deep-throated way that announcers of the '70's all sounded, and; [c] was scared shitless.

KTGR FM and AM in Columbia, Missouri was located in the Tiger Hotel, and was owned by the Tiger Broadcasting Company (this was probably because the University of Missouri at Columbia's football team was not known as "The Amalgam of Spoiled Kids From St. Louis and Dumb Hicks from Cape Girardeau"... but I digress). We had a control room the size of a phone booth, 2? to 3-minute records and a urinal on the second floor of the hotel, reachable only via a longgggg stairway. The owner/General Manager was an old man we rarely saw, who always wore a raincoat like a flasher from Central Casting. The only time I ever met him, when I was first hired, I stuck my hand out to shake his. He stood there, hand in recession, and stared around the back of my head. "It's TOO LONG!" he said, referring to my hair, ***(RM- thank God you clarified this, Russ)*** which for the time was hardly. "Oh, don't worry, sir, we'll see to it that he gets it cut", said my new program director. Welcome to Tiger Radio.

But it was that program director who proved to be the bete noir of... well, literally my entire career. "Mike Morgan" had acquired the job, essentially, by convincing the old man to fire his predecessor, a truly nice guy and talented broadcaster who had been commuting back-and-forth to St. Louis to train to get his First Class (engineering) License, and who as a result was too bushed to notice that this snarling young fella' had the knives out for him.

As fate often has it, this turned out to be a mitzvah for yours truly, because the sequence of musical chairs that followed on staff opened up a job for me... my first commercial Radio job! Zowie! Except, as I pointed out earlier, I totally sucked and had no clue what I was doing; and as I just explained, the program director was a complete shark of a human being.

In the early '70's, all the rules were much stricter in American Radio; there was mandatory News, mandatory station identification at precise clock times, mandatory precise transmitter readings, also at exact times and precise logging of commercial times (allegedly, those last two parts still exist. Except they really don't).

Now, try sticking a newbie into a "tight" Top 40 format, with all that mandatory Stuff going on, and you have a recipe for some serious disorder in the control room. Did I mention the cramped control room was in the WINDOW of the aforementioned Tiger Hotel? At the time, apparently, no one was worried that someone would shoot a disc jockey. Given my performance at the time, this still puzzles me.

So there I was, one Saturday afternoon, sucking eggs to beat the band as a putativeTop 40 DJ. One break after another, just new levels of baaaaad.

But then... oh. My. God. THE NEWS. "Rip and read" wire copy, ripped and edited during the short records I was playing. Coming up in another two and a half minutes; lasting five more, with an internal break for commercials at a precise time; weather coming out of it at a precise time; my "lockout", or outro, culminating in a legal "KTGR, FM and AM, Columbia!" I.D. and a time tone as punctuation. This required clockwork-like precision, executed live in real time. All this expected from a dweeb who couldn't read well, sound authoritative, or keep all the things I had to do straight in my mind; chewing gum was out of the question.

Record ends; newscast starts; I stumble over my stumbles. Sweat fills my shorts and my hands shake. I am, quite possibly, the worst excuse for a radio announcer in The United States Of America. It only seemed like an hour and a half; it was four minutes. And finally... saved by the bell! Or, more accurately, the 30-second recorded public service announcement. Now at last I could catch my breath, calm down ever so slightly, and regroup for that final 25 seconds of weather and then WHEW!

But it was not to be. As the "PSA" started to play, the old black telephone beneath the turntables started to ring. If you have spent 2 minutes in the Radio business, you know that phones do not actually "ring"; they "flash", because you're on the air, dummy. This one rang, loudly... it was the top secret "hot line", to be used only by the vaunted program director to inform the DJ at a strategic point in his broadcast that something needed his absolute, immediate attention.

"H-H-Hell---lo?" I offered. The buzzsaw-like growl of "Mike Morgan" greeted me on the other side of the hot line, roaring into my ear: "YOU... ARE...AN... ABORTION!" [Click.] 3... 2...1...

[Me:] "Ah... ahhhhh... AHHHH... K-K-KTGR w-w-weather..."

In the years since, I've somehow lost literally all my inhibitions on the air. I can talk to a cume of 4 million people with pure swagger, daring them (in my mind, anyway) to object to anything I say or do. And as I've explained, what I do in my sterling performance pays no tribute to decorum or convention, generally. But, true confessions time: I never, after that moment and to this very day, have ever fully believed that whatever came out of the radio while I was on, did not in actuality suck.

Bigtime.

The nightmares are ongoing, that the song isn't cued-up, or there is no song, or I don't have the news ready (there is no news, anymore), or I have nothing to say (not that this has ever actually happens anymore, either, but still...).

Angela Pause Smith

Reading school closings one snowy morning in the mid 90s at WTRY when the cart machine started smoking! I was a trooper and completed the list even as I was waving for someone to help - of course the smoke was thick, black and likely toxic.

Jon Knott

WTRY, early 70s. Contest was "Guess how many Mr. Pibb Twist Off bottlecaps will fit in this large hot air balloon". So I kept talking (off air) about how funny would it be if someone misspoke. Well, I guess I said it enough, because the very next time I played the contest, got the contestant online...I said "Can you guess how many Mr.Twibb Pissed Off bottle caps will fit in the balloon..." And, in the early 70s, you just didn't say "pissed off" on the radio. I've had plenty of the dreamy nightmare types too. I think we all have 'em!

Don Tandler

First job....At 16, I won $350 from 99X/New York, one of those scrambled song things. Went to the station to get my money, the PD's secretary happened to pass by, said "Hey! You're the kid who got that right! Thousands of people tried & got it wrong. You really know a lot about music! Want to work here?" "Uh...yeah"...... An hour later, I was compiling the "Spring 300", thousands of postcards with listener's votes, because no one else there wanted to do it. For a survey buff like myself, it was a dream come true.

First of many. Aug. 16, 1977, WTNL, 500-watt daytimer, Reidsville, Georgia. I'm PD; my best pal Keith Allen (now afternoon drive on B103/ Long Island) was doing his second? third? show ever professionally. I run in, yell "Elvis is dead, start playing his music!!" Keith played it, all right! At 78 RPM. It was awhile ...before we noticed. Poor Keith nearly wet his pants. Wish we were rolling tape..... :-) ***(RM- Keith you'll have to be in Book 2! Need to hear YOUR version of this one!)***

One station I worked at had an emergency button under the board that would alert the police. More than once, that button was accidentally brushed against. You can fill in the rest.....

Where I am today.....On air at three of Press Communications' stations in the Monmouth-Ocean, NJ market. Today, you have to make your dreams come true yourself.

Thom Whetson

I had something similar, only that it threw me on the floor. Later they showed me that if you touched a cable to the board and the right turntable it would burn a hole it the base that looked like an arcweld burn. Yay grounding.

Jamey Poss

I played Atlanta Rhythm Section's Imaginary Lover at 45 RPM, on purpose as it then sounds like Stevie Nicks singing, saying it was a new Fleetwood Mac tune and asking listeners what they thought! Other jocks kept getting requests for it for about 2 weeks before someone narrowed it down to my shift & I owned up to it! I actually got complimented for pulling such a great prank! ***(RM-Ingenious!)***

Bill Robert

First gig: WCCM-AM and WCGY-FM in Lawrence, MA. (Suburb just 30 miles north of Boston) Started out as an intern in high school. Was hired to do a weekend shift about a month before my High School graduation. Stayed there for over five years before another job came calling.

Back in my days as Ops Manager at WQRC-FM (Hyannis, MA) we had an annual Mother's Day promotion on the Cape Cod Dinner Train, which is basically a scenic train that runs around Cape Cod and a five-course gourmet meal is served on board. We gave away dinner packages on the train, where listeners could bring their mothers for a nice gourmet dinner on the train.

Dinner comes and goes, and all of a sudden the GSM and I notice that each table is getting a check. The GSM goes to investigate and is told that there was no alternate payment arrangement made for this trip on the train with the meals (which was BS because we traded it out and ran spots up the wazoo for the CCDT), and once the train leaves the station that is apparently it. So he and I are looking at each other in terror because we think we're about to face a mob of angry passengers. (BTW, I brought my Mom along, and she was having a good laugh at my embarrassment!)

GSM to the rescue! He whips out his Gold Card and picks up the tab for the whole freaking train! The next day, the two offices straighten out the issue and the GSMs Gold Card charges get reversed.

The best part of this: everyone on the train is thanking us for arranging the dinner and picking up the tab as well. We came off looking like heroes! (... See MoreWell, more so him because he whipped out our plastic savior, but people knew me from the station and came up to me as well. To this day, that guy is one of the coolest salespeople with whom I had ever worked (not just for that incident, he was overall a really decent guy). ***(RM- I'm sure there's a lot of women wondering the same thing right about now, ... where is this guy today, and is he married??)***

I've worked in both radio and TV, so I have double the nightmares about stuff going wrong. Either I'm on the air and the carts don't fire or I'm in Master Control and my live feeds are down and my playback servers are crashing. Somebody help me please!

Where I am today: I am Operations Manager at Wilmington Community Television, a local access Cable TV facility in Wilmington, MA. I left radio as a full-time career pursuit in December of 1998, after getting bounced from three gigs in that year (downsized twice for financial reasons, had "philospohical differences" the third time).

My gig here involves programming our three stations, overseeing and training our volunteer producers, and overseeing the technical aspects of the facility. In addition to the video production classes I teach here, I also do a radio broadcasting class (we do our own audio stream that runs during our "Community Bulletin Board" times and our members come in and do their own radio shows). I even got to build a radio studio as part of our new facility.

Do I miss being on the air? Hell, yeah. So I dabble in internet radio, doing my own stream through Live365. Most of the time it is just a standard web radio stream (which I program) but I can also go on from my home studio and do live programs.

Jonathan Weitz

How about editing a phone call on the shortcut for about 2 minutes, not realizing it's potted up. Every listener heard the call playing forward, in reverse, and has heard you say "OK, can you re-say this part." Yikes! Oddly enough, no one from management called me out on it. SSH! ***(RM- I'd like to introduce you to someone that if I have it MY way will be a part of my near future. Take a bow Jonathan..)***

Jeff Allen

Both Radio and TV as well for me. Back when we still had records and cart machines...I was reading the weather live...had one hand on the cart machine after starting the weather bed...then touched the turntable with the other to start it and ZAP!! I screamed like a little girl over the air. The city cop was listening and showed up to see if I was dead. Good times.

That little incident started my path to Engineering. Never had anyone shocked on my watch. LoL ***(RM- all readers here know, most engineers can't STAND me! Too much work for them! Glad to have you in my back pocket! I can blackmail you into doing stuff now!)***

My first "real" engineering job....the previous Engineer did not leave on good terms. Left a few wires connected to the mains in a dark cabinet under the Gates Yard Console with the ends stripped. The guy thought the GM would take over instead of hiring somebody, the GM used to take care of Engineering.

...or at a tiny TV station in Twin Falls, ID...leave for the bathroom and accidently put the camera on the air as the anchor is sitting there reading his script and picking his nose...the entire time your on the loo. I do miss TV.

Al Case

When I worked at WGLI in the mid 60s, there was live copy in the studio for the Sunset City Shopping Center. Within 24 hours it became a recorded spot. (I wonder why). BTW, I didn't do it..

The same thing happened to a dispatcher at the local fire/rescue alarm center when dispatching a call to the same place. The 1st time he dispatched it he very carefully pronounced every word. On his repeat he got careless and you know what happened.

Some years ago the name was changed to Sunset Plaza. I guess there were too many "accidents".

Russ Monroe

A competing station did a key contest to qualify to win a new Dodge. 104 Qualifiers won a key that they would bring to the big party at the end of the promotion. The listener with the key that worked, won the car. Day of the big party... broadcast live... EVERY key started the car. Gulp. ***(RM-Hahah! That's better than Lets Make a Deal!)***

... and then there was the morning sportscaster who had a little trouble reading a story about Mike Schmidt (Michael Jack Schmidt) but kept right on rolling along as if nothing had happened... while everyone else in the building lost it.

Anita Bonita

Two weeks after I arrived at WJRZ. Brand new to AudioVault; last station I had worked at on the air, we used RECORDS, for crying out loud. 7:30 News ... AudioVault begins to act funky during the spot before the traffic bed. So we hit traffic with no bed, and because the weather during that cast was ... "taped," I did it live.

Then came the conundrum: what the hell do we do next? No emergency CD in the control room, no engineer, no PD.

So here's what we did. I may not have known anything about AudioVault, but I worked for IBM between radio gigs and used to read DOS manuals back at Z-100. So I went about rebooting the on-air computer, while Jay and I talked continuously for about eight straight minutes—partially giving play-by-play, partially shooting the ... well, anything but the Breeze (the competition).

I got AudioVault back up and running just in time to hit traffic—with bed—at 7:43. ***(RM- THIS... is why I have her as my FRIEND. She's like SUPERWOMAN to me! A talent that is also the equivalent to a Female Engineer- that LIKES me!)***

AND...

WJRZ: "Ship Bottom... the most carefully-pronounced municipality on Long Beach Island!"

Tripp Rogers

I had a dream the other day...when out to grab breakfast, came back and the engineer dismantled the newsroom...I'm screaming at him "How will I do my newscasts?" and he says "Does anybody care?"...then I wake up sweating....

Another story...I'm working updates on 1050 ESPN on a Saturday in December of '06...I'm sitting in the newsroom watching a couple of games, when the WABC engineer comes over to me and informs me I have a sports update at halftime of the Seton Hall games is less than a minute...Raced into the studio with a couple pieces... of wire copy and a pen...jotted down the score and did a 1:30 off the top of my head using an AP scorecard, plugging the upcoming Knicks and Rangers games... the engineer was amazed I did that....so was I.

Frank Kingston Smith

First radio gig: WMAJ, State College, PA. I couldn't afford to work free for one of the Penn State stations, so somehow I got a job at one of the commercial stations in town. Ahh, there are host of great stories about that place. I started on the FM side, playing waltzes, gavottes and minuets. I graduated to the AM side playing the God shows early Sunday mornings and riding the board for the Prates baseball games on weekends.

I arrived at work (WABC) one morning to be greeted not only by "John the Nazi", our security guard, but by a well-turned out guy in a Brooks Brothers suit, at 5 AM. Guy showed me an ID which noted he was FBI. "NOW, what have I done?" I thought. Seems that some nut had burst into (I think) WLS Chicago, demanding to get on the air at WABC. He claimed to be wearing explosives. The FBI guy, and his FBI friend, sat with me in the studio until about 8:30, when his radio crackled that all was clear. Apparently the WLS thing had been resolved, and there was nothing in any newspaper about the incident. Jeez, all that trouble and the guy didn't even get any press.

Okay. I arrived at work to cover for Dan Ingram to be met by protesters. Seems that they represented the gypsy population and were protesting

our playing "Gypsies, Tramps and Thieves." It never ends.
Okay, a blunder with a quick recovery.

I was pretty good at over-producing the intro to any record. One morning I intro'd "Can't Help Myself" saying, "I need some HELP! Here are the Temps... and as Levi Stubbs sang "Oooo..." I realized I had screwed up ... good friends, the Four Tops!"

I have been told I should write a book, but if I did it would sell, like four copies.

Nightmare that actually happened... AT WVBF, the air studio was on the upper floor of a two-story building. I was (am) a fresh air fiend, so in springtime I would open a big window to my left. (I have to point out this was 1975-78... ...it's important.) A cold front was passing through and the breeze was coming up. Suddenly there was a BLAST of wind, and the arm on the turntable skittered across the record (yes) that was playing.

I opened the mike while laughing, and as I did, the entire window shade blew off the wall and onto me, the tables, the console, my coffee and my donut. I believe I was the first person in Boston to say, "oh, CRAP!" on the air! ***(RM- just giggling here, Frank! We ALL are!)***

After all those years of getting up with the cows, milking the public, I am REALLY glad that this page is up. ***(RM- :-)***

PS! I had a stalker before there were stalkers, 1975. WVBF. All I knew was her first name, and she had this cute little voice. She'd always open with, "Mr. Smith?" I talked with her for a couple minutes, but AM drive was busy. One Friday I had a recording session in Boston at 10, so I blew out as soon as I got off the air. About 9:15, I was told, a young blond girl showed up at the receptionist's desk and asked for me, very politely. When she was told I had left, she DEMANDED to see me, becoming a bit of a beast. When told that I really had gone, she began screaming and bit her arms until they bled. The local constabulary was summoned, and I never heard ANYTHING more about it. Strange.

I used to commute into WABC New York from Jersey via Rte 3, long before there was a "Meadowlands". A drive-in on the way was showing PLAY MISTY FOR ME. Someone with a great sense of continuity stole one of the letters, so the next night, it read LAY MISTY FOR ME. But, on the FOLLOWING night it read PAY MISTY FOR ME. Just grand.

And…

I had a CD skip on the air about 1992, working for WODS Boston (CBS). I was so angry, I took it out and broke it on the air. It took 6 months for them to replace it.

Another "Across the street, OMG!" We carried the Pittsburgh Pirates at my first station, WMAJ State College, PA. It was an "air" network; we received a signal from an FM station ... See Morewho got theirs third generation from the flagship, KDKA Pittsburgh. Bob Prince gave the system cue for the network station break at 8:57 PM, and turned play-by-play over to Jim Woods. I left the station and locked up to go pick up a sub sandwich across the street. I could hear Jim Woods, on the speaker over the front door, say, "Let's take a break here; this is the Pirates Baseball Network." While franticly fumbling for my key, I heard each of the five stations upstream do their IDs, right after: "Catch Rege Cordic, tomorrow morning, right here on KDKA Pittsburgh. By the time I made it to the studio, Jim was back into the fifth inning. ***(RM-I grew up in Pgh, and all radio jocks were my uncles!)***

I retired from radio in 1993, after 29 years. Stayed away from it until about 1996 in Phoenix, where we had moved, I had a talk station (which I hated) ask if I wanted to go back on the air. So I did, for two days. After which I walked into the GM's office (also the owner) and said: "This station sucks." Thus endeth my broadcasting career.

Ed Altman

I remember back in 1990, monitoring Casey Kasem's American Top-40 on a radio station just outside of Atlantic City. There was a 7-11 just next door. After years of being shipped to radio stations via vinyl, AT40 started using compact discs for their shows (with their own commercials already on them). Anyway, I saw I had about a 16 minute segment coming up. I started the segment, and was hungry. So i went over to the 7-11 to get some food. They had our station playing in the store, so I could hear what was airing. I remember Elton John's song "Sacrifice". I was just getting to the cash register to pay for the food, when I heard Elton singing "but it's no sacrifice, no sac-ri-fiiiiiiiiiiiiiiiiiiiiiiiiiiiiiiiiiii.." Yes, the AT40 CD was skipping!! OMG! This was on an early Sunday morning, and no one else was at the station. I never ran that fast across the parking lot to stop the skipping. I'll always remember that Elton John song because of that!!

Bill McCloutier

I'm GM of a cluster in upstate NY. The company just acquired another station and I'm unhappy with the morning show. I discuss it with the PD and he agrees we should make a move. I look over the talent contracts and all three have severance clauses, which, for me, make the move easier.

On Friday, right after the show finishes we call them into the conference room, tell them of our decision, remind them that they'll stay on the payroll during the severance period and then there is silence.

Finally, one of the three says, "Oh, great...on my birthday!"

For years afterward, the running joke was don't come to work on your birthday; you're apt to be fired.

Dave Hazzard

My first gig was in 1981 at a very small AM & FM owned by an old lady with one foot in the grave. The facility was located in a backwater I jokingly refer to as Moose Knuckle, North Dakota, but was actually in Upstate New York. Driving there from my home in Syracuse was much like traveling back into the 40s. Both the town and the facility were as archaic as the old lady and standards of decency did not allow for any reference to personal matters.

I'll never forget when the morning man got fired for mentioning magnesium as a stool softener in a human interest feature on daily supplements. Mind you, this was not presented comically in a 'morning zoo' format. The program was strictly public service and the host as dead-pan as Ben Stein.

In 1988 I was working a station in Fresno CA. and our console was low enough so that a simple office chair was used instead of a high-top. Anyway I'm reading a piece of promotional copy and during the last sentence I feel something warm on the crook between my upper arm and forearm. Soon as I cut the mic I looked down ...and there was a jokester colleague's skin flute on my sleeveless arm.

My absolute most horrific moment in radio came when I realized that corporate giants were systematically rendering the kind of radio I cut my teeth on irrelevant. ***(RM- WELL SAID!)***

These days I'm an aging Country Music jock in the #46 market, pathetically hanging on to an industry that long since passed him by. Thanks for asking!

John Tartaglione

This was from my time on KODZ Oldies 99.1 in Eugene, OR. I came in for my shift and the engineer had done some work on the studio monitors, they were off whether the mike was hot or not. So the studio was always in an unaccustomed silence and you had to be very conscious as to whether or not the mike was on.

Comes the time for a wx break, I key the mike, do my thing which goes from smooth to rocky to Oh My God, Shut-Up in no time flat. So I wrap it up, key off the mike turn away in disgust and mutter to myself, "Damn, tanked that f#(%ing break". No sooner had the words left my lips when I thought that I'd better check the mike. You guessed it, the lights were on and burning brightly for the masses. I keyed off & double checked it. I called home and asked "did you hear anything odd?" She said "no" & I relaxed, until the phone rang later and a buddy at another station said, "So, I heard you say f#%(ing x2 on the air today", I turned white, but my PD was out of town so it went under the radar. ***(RM-shows right there- even our spouses never listen to us! ON or OFF the air!)***

Dan Hardee

My first gig was as a Promotions intern for KVUU in Colorado Springs. I had to clean the vehicles, clean out the prize closet, go get McDonald's for my boss, and work for free!

My first pro gig was weekend overnights at WAYV in Atlantic City in the summer of '78...drove 3 hours each way to make my way to the roof of the Ritz Apartments overlooking the boardwalk. Watching the sun come up off the Ocean every morning is one of my lasting memories...as was working the Jerry Lewis Telethon from Resorts and having a listener hear me on the air Monday afternoon after hearing me at midnight---pledged $5!

Recently our Citadel cluster was holding a toy drive for a local non-profit. I was anchoring the mid-day news and the news director handed me a piece of copy for the drive 15 seconds before going on air. I began to read "They are still looking for donations of food, including Hams,

Turkeys, Hogs, Ostrich Meat, and Yack Meat. Anything you provide would be greatly appreciated."

Shortly thereafter, I received a lovely call from the PD asking why I was soliciting donations of Hogs.

Apparently, the news director had been typing the copy for someone else as a joke and got his stuff mixed up.

I'm now the Assistant Program Director for KVOR and KCSF at Citadel in Colorado Springs. I also serve as the Promotions Director for KKPK, and on-air swing talent for KKMG and KKPK. ***(RM-We see Citadel's getting their money's worth outta' YOU!)***

Mike Carter

Back in 2005, I was working for a station in eastern North Carolina, doing a remote from a car dealer, and as I was wrapping up, I thanked the FOLKS who stopped by to register for whatever we were giving away... well, instead of FOLKS, I'm sure you can imagine which F word came out. The AE that was working the remote with me turned white as a ghost, and I just tried to end the break as smoothly as possible. I never heard anything from anyone about it...and to make it funnier, this was just a month or two from me leaving to go to a CHRISTIAN radio station.

Hoyt Smith

Okay. First commercial radio gig in 1973 just north of San Diego in Carlsbad, CA at KARL-FM. Worked at the public radio station during the week and drove up to Carlsbad for the Sat & Sun midday or afternoon shows. The station was kind of a wreck. Bad, old Collins cart machines... not any sign of any real maintenance that I remember. The owner was trading/bartering for everything. This became apparent after they A: took away our AP news machine for non-payment of the bill and B: They couldn't pay us, but "how about some scrip for the seafood joint down the road for you and 50 of your closest friends". "Uhm, does that include my landlord?" I remember, too, that the logs would run out often around 2 or 3 in the afternoon and the traffic gal's daughter would come flying through the studio backdoor with more typewritten pages of the log. Was there for around a year and landed a job doing fill-in and weekends at KFMB-AM, a big personality AC station in the 70s.

Probably the morning about 5 years ago while taking a Commuter Quiz caller on KDFC/SF around 6:30 am I accidentally left my mic open under the music and after taking the caller's data and hanging up I plainly said into the mic he was an asshole (well, he was!). Heard from one listener via email a few days later that they were disappointed in me. At a listener party later a few listeners said that mistake won them over to my show. Casual, comfortable Classical, that's what we strive for. Oh yeah, my boss has never mentioned it to me, thank God.

Now, I am sooo lucky I even have a job as so many of the greats I know from my radio past are out to pasture; they are unemployed or under employed. I have been playing classical music at one of the last stations of it's kind in the USA, KDFC, San Francisco, doing the morning show since 2003. We were owned by the Mormon Church for many years, but now owned by Entercom. Great people to work for and if we had been sold to any other big radio group we would cease to exist. Entercom is committed to the music we play and to our voice for the thriving arts community in the SF Bay Area. It helps, too, that Entercom's founder, Joe Field is a classically trained violinist. We are commercial classical music's last, best hope in America. I'm hoping to work for another 8 years and retire, but one never knows, do one.

Good luck with the book. Need anymore assistance, let me know.

Bob Leonard

OK...here's another true one from my ABC Radio Network days. The studio phone rang and, as always, I started recording the call. It turns out the caller had his cell phone in his pocket and accidentally hit the number to call us....while the phone was still in his pocket. He was talking and said "shit", which was very clear. I edited the word out, assuming that since he called ME....accidental or not....it was implied permission to run the call. As I hit the playback button of the Voxpro...I inadvertantly hit the one that played back the unedited but....oops.....a weeks suspension without pay from a PD who had been trying to get rid of me for years...he didn't...I came back.

And...

First year of ABC Radio Network when it was Satellite Music Network (which, of course, Joe, you knew and loved as your place of employment). The wind chill is about 45 below zero. I have to be on the air by 5am.... it's 4:35 and I'm on my way from Frankfort to Mokena.....down a solitary road in the forest preserve. My door had been frozen open, so I stop to see if I can get it closed. I do and, of course, it's locked. I pick up a rock and miss the window from 5 feet away. Throw it again...it goes through the window, so I'm now driving to work in a 40 below wind chill sitting in glass with no window. After calling in a police report that a kid threw a rock through my window, I head into the studio to do the morning show and I am freezing and livid. In walks my news man (Daved Marsett) who plunks a rock on the console and says: "Here, you dropped your keys." That's when I realized how funny the whole thing was.......

And...

I don't even know where to begin.......WYSP, Philly....late 70's. Fox and Leonard are hosting the channel 13 (PBS) Halloween fundraiser, which was probably some bad horror film. As the camera came to us to interview Count Dracula, I noticed that my partner was nowhere to be found (I think he was in the bathroom)..., so I introduced myself and my guests, Count Dracula and The Invisible man....It was that day that I realized I could do improv, which has become a good part of my career off the air.........
(RM you are Hysterically funny!)

Blake Williams

(RM- I must preface this story, by saying this: I've edited every story in the book. Some were easy, some not. When I saw the length of this story, I thought to myself, "Oye Vey." The more I kept reading, the more intrigued I became! I didn't have to change a word, and I didn't want to CUT a word! THIS, is a wonderfully written story! So sit back and enjoy what Blake has brought us to the table!)

Dear Robin,

My interest in broadcasting began while in high school back in my home town of Casa Grande, Arizona. I began hanging around the only station in town which was a 1,000 watt daytime only station with studios located a few miles outside the city limits. I would ride my bike to the station and in retrospect, I'd be getting in the way a LOT! The equipment was nothing fancy. An old RCA transmitter about the size of 3 large refrigerators standing side by side with a giant evaporative cooler blowing air into the back of it. Next to the large monolith was a noisy Associated Press machine continuously typing away National and state news on a continuous ribbon of canary yellow paper. The studio had an old Gates Yard console (tubes) with 2 rumble master turntables and a rack of equipment that looked like props for a Frankenstein laboratory. The format was what was termed MOR ("Middle of the Road" or "Music of Robots"). The walls of the studio consisted of acoustic tiles now yellow with decades of cigarette smoke. There was a productions room but it had no lights in it and was used strictly for storage and a cat had made it her's so she could have kittens. The owner had a house trailer in the back of the building where he and his wife lived. They were nice people but not the kind of folks you would expect to find in a small town like ours. This was a farming community.

I had a friend who was 2 years older than me who had an afternoon shift everyday from 4 till sign off. "Jed" (now known as "Jed the Fish") was brilliant and his radio job was one way for him to cope with living in a town as slow paced as Casa Grande. Jed is now working for KROQ in Pasadena, California doing afternoon drive. He's been in that position for around 30 years now. It was his good nature that gave me my first shot at being a disc jockey on my 14th birthday. He let my do his show for 2 hours. That was it. I was bitten by the radio bug with no going back.

Soon I was working 2 days a week on weekends and being paid 100 dollars a month to do it. Soon the station sold and new owners dismissed Jed and hired me to work every day after school and weekends. I remember having to get my Social Security card and being paid a whole dollar and ninety cents per hour which was minimum wage. Soon, a new FM station was being built in town and that's when things began to get real interesting.

A rather young guy fresh out of the Air Force had a dream of building his own station somewhere. It turned out the only affordable license was for Casa Grande. Brett Eisley lived about 50 miles away just outside of Phoenix but chose to sell his home and relocate to our little town. One day before even breaking ground, Brett was talking with a lady in the towns post office who mentioned that one of her son's friends was 'good at radio'. Soon after, I got a call at home asking if I would meet in Phoenix to cut an audition tape. The new FM was to be consulted by a big name programmer in Phoenix at KDKB, the legacy FM rocker.

I really enjoyed sitting in a big city radio station's production studio with new equipment that actually worked with air conditioning, sound proofing and good microphones. The consultant liked my audition saying it was "conversational" and sounded many years more mature than my age. I was in!

In 1975, we began renovating an old house in the middle of town to make 2 state of the art studios complete with floating floors, solid oak doors, Neuman U-87 microphones, track lighting, JBL 4311 monitors.... It seemed like overkill for a town who's population was about 8000 people but considering the building was located one block away from the rail road tracks, the sound proofing was necessary. I loved being involved in the construction of this new 3 KW FM station. The format was Adult Contemporary with a few, pre-approved, album tracks sprinkled in. It was a good station to listen to even by today's standards. I was a Senior in high school when the station went on the air so I entered the Distributive Education Program (DECA) which allowed me to go to school in the morning and work for credit in the afternoon.

Fast-forward to 1984. I was working for an AM top 40 station that held it's own against all of the FM's in Tucson, Arizona.

For a couple years I was the afternoon jock on 13 K-Hit. Then I got the most unusual phone call I've ever received to that point. The voice thick, with a British accent. The gentleman said his name was Paul Russling and he was calling from Fort Lauderdale, Florida. He was there outfitting a ship with a 50,000 watt transmitter, studios, generators and everything needed to launch an off-shore radio station. He had heard that I had an engineering background as well as programming and on-air experience.

After a short conversation I was in the studio putting together an audition tape and typing up a resume. This package went to an office in New York City to a company that was to manage the radio operation from the Unites States even though the programming was to be aimed at a British audience. It worked. I was hired as the Chief Radio Engineer and an announcer. My life as a pirate was soon to begin.

I met up with the ship, the "M V Communicator," in Ireland. Immediately I was struck by the lack of a transmit tower. I heard rumblings that this operation was to employ large helium filled advertising balloons to hold up the antenna wire. Sure enough, welded to the deck, there were dozens of huge bottles of helium. The balloon antenna idea worked for Radio Marti located in south Florida transmitting to Cuba. There is a slight difference between the weather in Florida and the weather on the North Sea. Tethering a huge advertising balloon 411 feet in the air in gale force winds turned out to not be such a great idea. I think we got to use the balloon antenna for a total of 11 hours before the winds took the balloons away. Needless to say, another method had to be found to get a signal into the air.

To this day I'm not completely sure what was going on between the powers that be after the balloons proved to be impossible. All I know is everyone was taken off the ship except for the captain, another engineer and me. They all went for "shore leave". Our task was to somehow, with what we had on the boat, to get a signal on the air and run 24 hours of test transmissions. The other engineer, Johnny Lewis, a British National and real great guy, and I put our heads together and rigged a very strange looking antenna together running the whole length of the ship. We needed a special coil to help match the antenna to the transmitter.

Improvising, we came up with a way to wind copper wire around 4 florescent light tubes held in place by foam packing material. Somehow, we were able to get 5000 watts out of the transmitter, into this strange looking antenna and we had a signal. It wasn't a powerful signal but none the less, we had a signal strong enough for the money backer to receive near his home in Ireland.

During the 24 hour test transmission we played mainly Beatles songs and a few from the Rolling stones. The sea was rough, we were exhausted but yet the transmission went on for the required 24 hours. Everything held. Once we were finished I remember sleeping for 12 hours solid.

Days went by and we heard nothing form anyone. Weeks went by and still, we heard nothing. I had drawn the conclusion that the whole project was going to be called off. Remember, no phones. All communications had to be cryptic and short over the marine VHS radio.

Finally, after a couple months of being stranded on board the Communicator, a tender came with all of the other crew and disc jockeys.

I could no longer stay on the ship. I knew that it was going to be a long time before a tower was constructed on the ship so it was going to be a couple of months before we could do what we were brought out there to do. When I jumped onto the tender, I had every intention of somehow flying back home to solid ground and real towers. Once back on land in England, I settled down at a friends house in Whistible, Kent.

A few days after being on land I had an opportunity to meet the folks from Radio Caroline, the world's most famous pirate station operating since Easter, 1964. With the sinking of their ship the Mi Amigo a few years prior, Caroline had been silenced. Oddly enough they just came back on the air a few days before the Communicator dropped anchor 2 miles away in a place known as the Knock Deep, 3 miles off the coast of England in international waters.

Caroline had a new ship and a 300 foot tower for their 50,000 watt signal. Both ships could see each other from the anchor positions. While meeting with the folks from Radio Caroline, I was offered a job. They heard me on the air during the test transmissions and they were also looking for an Engineer. As I was having difficulty communicating with the New York office running the American operation and I had serious doubts about even possibly being paid, I took the job on Caroline. What an experience this was to be! It was 1984, Ronald Regan was President; Margaret Thatcher was Prime Minister. Princess Diana was getting married to Prince Charles. Interesting times.

Today, my life is much simpler and far more stable. I live in the high deserts of New Mexico on 5 acres with my horses, llamas and dogs.

Thanks to the internet I am producing radio and TV commercials for stations around the world. I also produce a 3 hour program each week for www.replayradio.net, a station based in the UK. I've been fortunate enough to participate in a few feature films made here in New Mexico.

Besides acting, I've also gotten into working on sound tracks as a sound designer. More exciting things are on the horizon but I do not wish to jinx them by saying anything at this time.

Should you have any more questions, Please ask.

Blake

John Hutchinson

How about when you are in the studio alone, you need to go to the toilet, and you put on a record in the hope that you can get back before it finishes. Guess the rest. It finishes before you do!

Matt Shwind

Robin - thanks for doing the book! Great idea. I'll definitely buy a copy and give one to all my fan. ***(RM- So cute!)***

After 10 years of radio I became a realtor - still selling houses today.

My original story from my first radio station gig in summer of 1979 at WHBF-AM "Country Sunshine" (Yuk)... I was 19 years old doing weekend overnights at WHBF in the Quad Cities (Rock Island, IL and Davenport, IA). It's 2am and I'm on the air - the only one in the station and it's in the middle of the biggest summer thunderstorm EVER. Every 5 minutes I'd run to the news room and back with the weather service tornado and thunderstorm warnings for every county within 40 miles. After reading this over for about the 5th time, I finally summed it up by frantically shouting into the mic, "if you're within the sound of my voice, take cover NOW!"

Christina Kelly

Okay, I accidentally yelled the F bomb into a live mic. Lucky for me it was only potted up to about 40--and I had one of the best PD's in history. No drama. But our consultant (also one of the best) did come running down the hall and while bursting through the studio door (frantically and wide eyed) he shouted, "where's the truck?" ***(RM-This one had me laughing for DAYS!)***

John Willyard

1. College radio, KCMA-FM San Francisco (doesn't exist now); first paying radio gig was KEAR-FM San Francisco and its radio network facilities.

2. Prepping/editing the 2:55 AM newscast on a Monday middle of the night, putting my head down in front of the control board at 2:40AM, waking up at 3:04 AM — and NO one called!

3. I'm sitting in my Marietta, GA home studio downstairs voicing TV affiliates, Country radio stations, cable including CNN, GAC and HDNet, narrations, commercials, voice of the CMA Awards since 1996 — whatever I do that's somehow keeping me too busy to golf these days. But I know I'm very blessed — and I am grateful!

Lisa Allen

I had been on the air for less than 2 years. I had just started working weekend nights at WIRK/ West Palm Beach and had been there a few weeks when during a Saturday night shift, someone took control of the transmitter by remote and had a 10 minute or so pre-produced thing which contained all kinds of profanity. I was so green that I didn't have any idea what to do (I should have turned off the transmitter). I just sat there and listened. Just as it was ending, our morning guy came in the building after finishing a bar gig. I was shaking and could not go back on the air. He opened up the mic, explained what had happened and offered an apology to our listeners.

Kristin LaBar

"My first radio gig was officially on MAX 106.3 WHCY FM.....in 2006/2007. Before that, I was on 90.3 WESS FM at the age of 15 and 16 (1992/93)... before I could drive a car! I've been a radio geek for a long time! LOL. I also "interned" over at 93.5 WSBG when I was 14, with a great DJ named Mick Ryder who let me help out and learn the ropes of radio! In December 2005 I started an internship at B104 WAEB FM in Allentown, PA and got to hang with Scott Lowe in the studio every week while he did his night show. He knew I had done this all before, so we jumped right in and started making demo's to try to land me a radio gig! Scott also taught me a lot more about radio! So, as I continued to work at WAEB FM in promotions, off I went to MAX 106.3!

Then after 5 months....it happened! My dream came true as I got a part time gig on Fresh 102.7 in New York City! I'm now hosting the Saturday Night 90's every weekend from 7 to midnight. My radio career so far also includes stops at 97 WBHT and Magic 93 in Wilkes Barre, PA."

I was on the air one night, and I had the radio in the other room turned up a little so I could hear my break after I did it...(we are on a delay), SO...I did the weather, started an REM song, and it happened to be a long intro, and a quiet one...so, I accidentally left my mic on and up...after I hear my weather break play...being as critical of my breaks as I am sometimes...I let out a big "EWWWWW"...well, I'm sure everyone listening heard that one! LOL. Now I check that the mic light is OFF before I say ANYTHING! :-) ***(RM-Always check your board Kristin, when you follow another jock! –insider)***

James Clark

Ever ordered pizza and end up in the ER? I was working during the ACC Basketball Tournament and decided to order pizza. So here I am waiting when the pizza guy pulls up, drops the pizza off backs up and plows into the paper and mailbox. So during the break I went outside to get the boxes up and take the stobs out of the ground. Took the first one out no problem, Second one was easy but as I was tossing it to the side the edge caught me cut me wide open from my palm to the tip of my thumb. So here I am bleeding running in and called my boss and told him about it, and then went to the ER to get my thumb sewed. Doctor looked at it and said: "It's Not That bad." Here I am bleeding still and he says, "Its Not that bad?!?" I finally get a band-aid, and head back to the studio. Finished my shift, put all my carts away... walked out the door, got in my car, backed up, and then you hear, "POP" I just ran over the stupid stob that cut my thumb. Some jerk put it back in the ground. So yeah...
AND...

I was doing production for the local AM station when I had to walk out to my car to get my lunch. Knowing that radio station food sucks especially from a little snack packer thing, went outside. Three minutes later a crack of lightning went across the sky.

So ...here I was running back to the door, and the door was locked. I went around the back and found out that the news peeps left the door unlocked, "SUCCESS!" I thought when another lightning bolt hit the sky.

As soon as I got in, I hear a sudden pop and thunder, and the entire studio went dark, and came back up. EAS sounds off that we are now under a Tornado warning. After that, we go dead air. Lightning had run into the stations main Audio Vault server and fried a mini-micro chip and that was all she wrote.

So yeah.. Dodging bullets is always something to go by. LOL ***(RM- So glad you're on the mend James! We were all rooting for you throughout your recent accident!)***

Kris Pierce

Wish it had been a dream—I was reading a news story about a man streaking in protest of some political something or other, and instead of reporting that he was "apprehended, but not arrested," I boldly stated he was apprehended, but not erected... My program director/morning show co-host dropped all the carts he was pulling!

Jack Shell

First gig: WXCL-FM/Peoria (overnights)

Love the one where I'm in a foreign radio studio, and it's usually my competitor doing an air shift and all hell breaks loose. That used to be a recurring dream before I went to work at my current station... now that dream is reality, however I know now how to use all the equipment.

Or….

1.) There is a tornado, and I'm still on the air acting as if nothing is going wrong outside of my control room, yet nothing works right, and the roof gets ripped off part of the radio station, and I'm still on the air oblivious to the tragedy that just occurred.

2.) There is the one where I am doing my show on the competing station, and have no idea what the hell I am doing there.

3.) My least favorite is the one where I am working for that one station I really hated, for that one boss I really disliked. I always wake up in cold sweats from that one.

Current gig: WKDF/Nashville (afternoons)

Stace Karge

Okay...here's mine...still haunts me...

Its the Oakland World Series. I'm running live-assist on a station programmed for a demo that didn't get the memo that they were supposed to have been dead 15 years ago (anyone remember stardust memories?)

So I come out of the network news and the out cue is something like "So far... 17 confirmed dead." and the first spot I hit says "If you haven't made out your will...now's the time...the American Will Kit..."

I literally started screaming at the cart machine "NOOOOOOO!!! I DIDN"T JUST DO THAT!"

On the other hand...it could have been the most effective marketing positioning they ever had...I've always wondered... ***(RM- I think...I'm with ya' there?)***

:) Stace

Vinny Marino

I just had one over the summer. I'm doing the morning show on classical WQXR in NYC. At 6:35, right after the news, I start the next piece and suddenly all power in the studio goes out. Xmitter is fine, but NOTHING in the studio is on, including the lights! Dead air for 3 minutes until the emergency backup feed kicks in which it does. I call the Chief Engineer and he has me go into a room with electrical equipment that looks like something out of "The China Syndrome." I literally have to PULL A CRANK to restore power to the studio! Had to reboot the Scott System and the Internet feed. Since classical pieces are rather long, no spots were lost, so no make goods!

Tom Rocky Pantaleo

One time at WRCN , I had to use the men's room, which was in another part of the building, OUTSIDE of the station, so when you leave the station, best prop the door open, because NO I didn't have a key, so I'm playing a 35 minute music block, and think I have the time. I walk out the door, and place a coffee cup in the door jam, and walk away, only to hear the door close and slide the cup out of the way....HOLY HORSE SHIT!!! I couldn't find my way into the station, and by now I have 10 minutes before ENCO stops.......I spied a fire extinguisher in the hall way and beat the shit out of the doorknob, broke it off and ran into the studio with the last LONG WAY HOME from Supertramp fading.....Nobody knew anything, but the PD called and wanted to know why I was out of breath! This was a Sunday and to have a guy come and fix the door was 800 bucks!!! ***(Rm- Rocky, this is Priceless!)***

Eddie Stambaugh

I do weekend/ fillin's at a country AM-jammer. We're live until 9pm and then set the sweeper in motion till 6am...last weekend, I do the usual triple check all the doors, do the trash, shut down the production studios...

yada, yada....set the alarm, walk out the door jump in my car listening to the last song I loaded in fade to black...AAAAH...some bonehead forgot to click the “Auto” button... (of course I was the only bonehead there at the time) so, in a mad scramble to flip my car around, dig in the dark for my station keys only to realize....I didn’t have them...anywhere...so I blaze my way up to the station door only to see my station keys just a shinin’ in the security light about 3 feet on the other side of the door...Doooh!!! ...Fortunatly I have a spare key at home only 10 minutes away. My wife, not happily mind you, came to the rescue!!! Glad to hear I’m not the only one with self induced “sweeper” trauma!!!!

Eric Rhoads

I was in a local singing group called ‘Sing Out Fort Wayne’. We did lots of "Up with People" shows and lots of civic projects. On this particular day in 1969 at the age of 14, I met a guy named Charlie Willer and we were on a team to break up the ice on a river in Fort Wayne to prevent flooding. We had been working at it all day and he said that he had to go. "You can't just leave," said I, "we're not done." His response, "I have to, I've got to go to my job on the radio station."

So I went with him, watched him do his show, and a week later I was on this same carrier current station at the Indiana Institute of Technology. WITB 550 am. I did that for a couple of years until I landed a church tape job at WLYV in Fort Wayne.

My career then took me to WYYY Kalamazoo and I was on the original air staff August 3, 1973 of Y100 in Miami, later on WMYQ which became 96X, and then WQAM in Miami. I went to my first programming gig at KIOY (K104) in Fresno where I was number one in the first book.

I then started a consulting company with Jerry Clifton in a company we called ‘New World Communications’. We were looking though the dictionary for a company name.... it was the New World dictionary... thus the name. We consulted about 25 stations at the time until I bought one of our client stations KEYY in Provo, Utah and the FM, which I named KLRZ (Color 95) in Salt Lake. I also got the license on 106.7 FM in New Orleans. I sold everything in 1986 and started the Giant Boom Box company, which was a promotional studio that looked like a radio. When I was advertising I encountered the Pulse of Radio, ended up buying it and turning it into Radio Ink.

One time I was on the air in Miami at 96x when Jerry Clifton our PD said, "we've got to win this book against Y100, so I want you to push the limits on outrageous, just don't loose the license." I told the jox that day to say I had been fired. When it was time for my shift and Steve Rivers was sitting in my slot. When he took a bathroom break (which he told his listeners) I supposedly sneaked in, took over the station, nailed the door shut and played one novelty song over and over for about two hours. Great radio theater. During one break I had my headphones on, was about to open my mic when I heard a voice say, "Don't touch that button." I turned around and saw two Miami Beach cops with their guns drawn and pointing at me. Suddenly there was dead air. In hindsight I wish I had turned the mic on and had the whole incident on the air. The cops cuffed me, read me my rights and started to carry me out of the building. (They got in when Steve Rivers left. He walked out the door and they threw him on the car, cuffed him, and arrested him but first made him open the door to the station.) As they were taking me out I said, "You know it's really a shame. Here you are just doing your job and you're the ones who will end up in prison". The cops said, "What do you mean?" I then told them I was licensed by the FCC to run the station and that by forcing me away from the controls they were violating Federal law. They asked who could confirm this, so I said call Carl Como the General Manager. I'll dial it for you. So they uncuffed me and I dialed Jerry Cliftons number, woke him up and said, "Carl, the police are here and want to talk to you." He immediately played along, I told him what I had told them, he confirmed it, and they set me and Rivers free.

The following day the police chief called Carl Como, the GM, and made the station run an apology every 15 minutes for two solid weeks. The apology said, "we would like to apologize for stunts performed by Eric Rhoads and Steve Rivers. We assure you it will never happen again." It was the best thing that could have happened to me. Suddenly everyone in town knew my name and wanted to know what I had done. The station's ratings soared for that rating book and we came within 1/10 point behind Y100. ***(RM-I read your stuff Eric, and I totally love knowing there's another 'outside the box thinker' like ME out there! Makes me feel less lonely!)***

Thanks for the opportunity Robin.
Eric

Charlie Mills

Robin- First real gig…

April 1973- WDVL Vineland, New Jersey.
I was hired to play country music on WDVL-AM, a 500-watt daytime station. The signal actually could be heard on the other side of the station parking lot! A blow dryer had more wattage! The first day on the job was a nervous one for me. I sat behind the board before turning the transmitter on to become familiar with the controls.

I inserted the "National Anthem" sign-on cart, hit the "on" button for the transmitter, and waited a few seconds before pushing the start button for the anthem cart. Then I realized I needed to cue up the first record. I pulled out an old Elvis Presley tune called "Twenty Days and Twenty Nights" and placed it on the old prehistoric 1950's style turntable. Then, as I went to cue the record I was astonished to discover the lack of a needle! Quickly I turned to the other turntable - again NO NEEDLE! I ran into the FM studio and asked the DJ where WDVL stored turntable needles. He replied "Didn't they tell you?" "Tell me what?" I said. He replied, "You have to bring your own needles!"

This is a true story! Luckily, the FM DJ, Tom Russell was kind enough to lend me an extra needle he kept in his locker. Kind of makes you appreciate today's technology. ***(RM- Nowadays, if someone said that, we'd be thinkin' heroin! This is stuff you just can't make UP! Too funny!)***

My worst radio nightmare?

Two actually. One was when I worked the overnight shift (1979-1983) at WPEN in Philadelphia. The building was like a fortress and I was convinced that one morning I would walk out to the parking lot after my shift and discover that a nuclear war had taken place and I was the last person on earth!

The other - Being on air, behind the board, with a bunch of well-stocked cart racks behind me. As a song ends on the air, I hit the button for the next cart, and it jams. I have no other carts in reach, and I have the mic on and can't reach the carts behind me because the chair is stuck in position - and for eternity I am left to ad-lib for fear of having any dead air!

I left radio in 2000 after being an Air Talent, Production Director and Program Director for various stations including WPEN in Philadelphia, and WSUN-FM in Tampa Bay. I currently own a voice over/copy writing business in Tampa Bay and consult a video production business in New Jersey.

http://millsmedia.net

Thanks for the invite and possible inclusion in your book!
Good luck with the project!

Ted David

First gig was at WABC Radio. It was 1967 and I was 18. I was a page. The jocks and staff announcers went out on strike. I stepped up and was on the air for 16 weeks for $1.25 and hour. They came back and I moved down a few markets to WLIX, Islip.

Worst nightmare was my first shift at WPIX-FM. It was an overnight shift and I was all of 21. I was so nervous that for the first two hours I had the mic turned around and was talking into the XLR connector.

Oh, and on that board the master gain and the monitor gain pots looked the same so each time I dropped the studio monitor to cue a record or hear myself think, I actually was turning the master down creating dead air.

I guess no one heard or cared because I was asked back for the next weekend and that started a six year relationship that culminated in 2-1/2 years in PM drive there. ***(RM- I STILL do that stuff TODAY!- thus my earlier comment about engineers not liking me too much...)***

Where am I now? I recently retired from a 20 run at CNBC and went to work at 1010WINS as a free-lance anchor. It's a venerable station but not a job for someone looking to slow down.

After 15 months I just retired from that job... and am now exclusive to News12 Long Island as a free-lance anchor.

Jim Quinton

Hey Robin,

I started in radio when I was 16, this was back when you had to take an exam at the FCC in Atlanta to get your Third Class Liscense with Broadcast Endorsement. I stopped by the local station, WYYZ AM 1490 here in Jasper, GA and talked with the station manager, N.J. Tippens, and he told me if I came back with a license he'd hire me. I'm SURE he thought he'd never see me again. WRONG LOL! I started out doing the Sunday Morning Gospel show, that branched to include Saturday afternoon country and then night time Top 40 during the week. I had everything happen to me that could ever happen at that station.

I had my news copy set on fire while I was reading it live on-air, had copies of Playboy's centerfold help up in front of me during the obituary column reading, even had to stick my hand in a live running transmitter with 1,000 watts and jiggle a tube that had gone out due to there being standing water in the transmitter due to a leak in the ceiling above it! ***(RM-this paints SUCH a visual! Theater of the mind!)***

These days some 31 years later I'm back in radio and working for a nice country FM station about 40 miles from here at Mountain Country Radio, WPPL-FM in Blue Ridge, GA doing 10 to 2 and programming the music. I moved to Nashville and did record promotion for a while and managed a couple of cd stores but never lost the love for country radio. Now I'm back and happier than I've ever been!

Hope that helps out and you make a TON of money for your charity! Let me know if I can help out in any way.

Jim

ROCKWELL AWARD

Kipper McGee

First "real" radio station: an AM/FM combo known as "Funlovin'15Q" (later Q102) licensed to Two Rivers, WI. I was managing my campus radio station (non curricular—we didn't even have a faculty advisor), so I started visiting local stations to learn from them. Eventually, I ended up working for all of them, and 15-Q was the first. Under GM "the Real" Howard Johnson and PD Denny Carpenter, I learned the valuable lesson that major market was an attitude, not a location!

My biggest distaster was at my first PD gig, when the 1800' tower of 100,000 watt FM108/Waterloo, IA collapsed during an ice storm. We were broadcasting live from a Police concert when the station started going off the air, then coming back on. Turns out, it was an ice storm, and although the tower had de-icers, they guy wires did not, and the wind was pulling the tower and STL out of alignment, thus knocking the station off the air. Brutal without a backup, but the story has a happy ending which I'm happy to share sometime if you'd like. (Both of these GM's were present when I accepted the Rockwell Award at last year's Conclave!) ***(RM- I was there tooooo! Congrats on that one Kipper!)***

I am now consulting a number of radio vendors, stations, and individual show on the concept of "Brandwidth" (using ALL platforms—AM/FM/ digital, etc.) for fun and profit!

Good luck on your project. Please let me know if you need anything further.

Best, Kipper

Tom Housley

Hi Robin,

Sounds like a fun idea for a book - I'd buy it!

OK, my time as a DJ was brief; I did radio in college but graduated to behind-the-scenes work in TV after I graduated from Ohio University.

The best "oh shit" story I had was this: I had a communications class taught by the Dean of the communication department. This occurred the day I took my final exam - a couple of hours after finishing the final, the Dean & his son were touring the WOUB-FM studios during my air shift. I saw them outside the window, pushing the button to listen in (which I hated - always made me feel like I was on display in a zoo). I waved them into the studio (of course). The Dean was standing behind the turntable that was on-air (yes vinyl!) and I was pointing out the news booth & the AM studio opposite me. Well, the Dean leans over to look just as I'm putting on my headphones - and his suit coat knocks the tone arm completely off of the album! Thinking of my 'yet-to-graded' exam, all I could do was give the old "uh, that's OK, don't worry about it" speech as

I scrambled to cover the dead air.

Then there was the time at the carrier-current station where I was music director that my naive friend played the Rolling Stones "Some Girls" song without knowing it contained not only a racial slur but the dreaded f-word. I found out about THAT the next day - but I'm sure that no one heard that one b/c we were not only on carrier current, but I think the only students that knew were even on the air were the ones that worked there....

I've been in local TV ever since graduation - I'm now a commercial writer/ producer for WHIO-TV in Dayton Ohio.

Hope this helps, ***(RM-Tom, Adorable!)***

Tom Housley

Dutch Guy

Hi Robin,How are you doing? I think it's a great idea, to write a book about DJ Nightmares. God knows we have plenty of them :)

Btw, when is your other book gonna' be released? ***(RM- after this one. I'll tell you more at the end of the book. This way I grab ya' ALL!)***

Well, I'm gonna' share a story as well. Don't know if you're gonna' use it, because it's a bit grouse. If you're gonna' publish it, use my on-air name: 'The Dutch Guy', thanks :)

It must have been 1996 or 1997 when I encountered my worst Disc Jockey Nightmare. I had a daily show, 7pm-10pm and had been prepping all day. Even landed an interview by phone with a local singer/songwriter, who was known of not doing interviews that much. This was unique, I was thrilled!

Started the show as usual and everything went just fine. The interview was scheduled for 8:30pm. Had some great questions for him, not those standard questions, but I really wanted to know more about the person behind all the fame.

8:30pm, record ended and I introduced the artist. After a short introduction, while I was firing away the first question, my stomach makes a 360 flip and I was about to vomit. Out of the blue! I could barely finish my first question. While the artist starts to answer the question, I close the mic, grab the nearby bin and there we go.

During the whole interview, after a question, I had to vomit and to be honest, I haven't heard one single word of what he said. If I hadn't air checked the show, I wouldn't have known a thing about the person behind the fame.

Lucky for me, the artist never suspected a thing and thought it just went great. As for me, I was sicker then a dog for the rest of the show, felt bad for the Disc Jockey who had to do his show after me, went to the doctor's office the next day and had a bad case of food poison.

I never ever hope to re-live that again. There's nothing worse then doing a radio show, say something on the air and just pray you don't have to vomit, while talking.

Started at several pirate radio stations, that was the only way to get some air-time, back in those days over here in The Netherlands. We didn't have commercial radio back then.

At the moment I'm a freelance jock, trying to get his big break on an US radio station. Including my sillge Dutch accent :)

Again Robin, if you don't wanna' publish it, I understand. But you asked for the worst nightmare, and this was mine :) ***(RM-Lex, I'd never ask without following through! Your story is WONDERFUL!)***

Have a great day!

All the best,
Lex

Rob Bennett

Hi my name is Rob Bennett, decided radio was what I wanted to do with my life when I was 12. My first station was WCLX in Boyne City MI. in 1981. I worked there 2 weeks did not get what they promised me pay wise and left, went to Ann Arbor MI to WNRS/WIQB... worked there 4 months and got picked up by WWWW Detroit MI.

While working at WLTI in Detroit, which was on the Transtar Sattelite with the exception of mornings and afternoons, and I was the afternoon jock. We had board ops that would plug in local spots and weather breaks for the other shifts. One Friday when I was getting ready to head out for a Great weekend I went in to record that evenings weather...I was recording direct to cart to speed things up...we had big police lights that said we were recording outside the studio doors but I guess the production director didn't see it and came in and messed up the recording so I turned to him pointing and yelled DICK! DICK! DICK! After he slunk out of the studio I put the tape in erase mode in the deck and after it recycled

I recorded the weather. That night about 9:30 the board op called me laughing hysterically and played the cart that had JUST AIRED... that went like this ... " W-Light weather calls for DICK DICK DICK!" Then while still laughing she said, "do you want me to save that for you?" I said "NO pull all the tape out and throw the cart away!" Luckily no one called to complain ...WHEW! ***(RM- I think we'd all agree on "LOLOLOL!")***

Currently I am working in Oskosh WI at WPKR 99 5 The Wolf doing mornings part of Rob and Louise in The Morning our website is thewolf.fm.

Thank You
Rob Bennett

Scott Baily

Robin,

What does FB stand for? ***(RM-adorable here!)*** Remember, I'm a small AM Radio Station owner, just outside Nashville, almost in the rural area. Being the only employee is hard, and with the economy the way it is, I can't afford to hire anybody to help me. I am also semi-disable, but I get the job done, thanks to the internet!

I contract things out like my local weather, local news, and local commercials being doing for the "mom & pop" businesses I have on. They can't afford to spend much, I do go to get between 50-100 dollars a month out of them, but we make it. The company that does my local news and weather trades with me. I run spots for them for their services.

I doubt I fit in your group. My station is on satellite with Citadel Media, playing Hot AC, Monday thru Saturday, then all Gospel Music on Sundays. Like I said, I'm just a little fish in this big ocean! ***(RM-we are all fish in the same sea... xoxox)***

Best Regards,
Scott Bailey

Christopher David Laursen

Rebooting NexGen during a blink 182 song. ***(RM-giggling here!)***

Chet Atkins

Jay Noble

First Radio Job, 1967 Armed Forces Thailand Network, U-Tapao Airfield, Thailand. Armed Forces Radio & Television Service.

First commercial radio job, KIAK-AM, Fairbanks, AK. 1972.

Worse Moment. KIAK-AM, Fairbanks, AK. 1972.

I was interviewing Kenny Price. I have to set this up a bit. I'm about 5'5" on a good day. Kenny Price was about 6' 6" and weighed in at around 350. KIAK was a brand new station and the first Country station in Fairbanks. Kenny was one of the first artists to come by. He was appearing at a local club. Kenny said, "why don't we play my new record." He had just released his version of "Sea Of Heartbreak." I said, "Kenny, we don't have your new record yet." He said, "Can I make a long distance call on your phone?" I said "Sure." Kenny dials a number and says, "Chet Atkins please"... waits a beat, "Chet, this is Kenny. I'm up in Fairbanks... that's right, I'm going to be at Rex Gavin's Silver Spur tonight. Chet, I'm at Kayak Radio talking to Big Country John... Chet... they don't have my record. Who used to be our record rep in Seattle?" Kenny then told me we'd have a case of his LP's the next day. He said, "The music business is cyclical... what goes around comes around. If you don't have my records you can't play it. If you can't play it, the listeners can't hear it. If they can't hear it, they can't buy it. If they don't buy it, I don't eat... and you can see by the size of me boy, I like to eat."

Current position,Program Director and Affiliate Relations Manager, "Moby In The Morning Radio Network." ***(RM- Wonderful story!!! We LOVE this! Best of luck with Moby)***

Dave Scott

I may not qualify since I'm not "on the air" except by recordings, but I have been. ***(RM-Dave? Paleeeeze!)***

To answer your questions:

1. First commercial radio gig - WPAG, Ann Arbor, MI, city council reporter. Was a DJ on the student station at the University of Michigan before that. Followed by part time work while a U. of M. student at most of the commercial stations in Ann Arbor, Ypsilanti & Saline, distant suburbs of Detroit.

2. Worst radio gig - all night DJ at 50KW top 40 WPTR, Albany, NY. Was repeatedly promised day shifts when there weren't any openings, but a PD change followed by a day shift opening that I didn't get.
Took vacation and never came back.

Found a job at KSHE, St. Louis.

3. Current on air - weather announcer on about 50 stations across the US on my UnattendedWeather.com service.

4. Other radio jobs, night top 40 jock at WQOK, Greenville, SC with 57% Pulse Rating, PD/VP/jock at KIRL, St. Louis

5. Syndicated The Z Format, The AC Format, The EZ Format, Super Country and others at Century 21 Programming, Dallas.

I programmed and was voice on 400 stations 1975-85.

In 1985 invented GoldDiscs and Hit Disc CD libraries.

I was CEO of Century 21 and TM Century through 1992.

6. Started, owned and was CEO Scott Studios, Dallas, 1992-2004

7. Currently own and CEO of Scott Traffic, traffic/billing software for radio. Also, consult Enco Systems (automation vendor in Detroit).

Larry Miller

First Radio gig- calls WEVD AM & FM New York City 1960 (I was 17)

2. WORST radio experience- calls KSADI St. Louis, MO. The owner was a true pervert...very nasty domineering man...sicker than me so you know that says a lot.... ***(RM- to all that don't know Larry? To know him, is to LOVE him!)***

3. Where you're working today. I own www.Backofthenet.com ***(RM- if you love Soccer- you should all check out his site!)***

Mike McCann

1st radio gig - weekend/swing DJ at WLNA Peekskill

Worst radio nightmare - June 3rd, 2005 - being told to join a conference call taking place in 45 minutes during which CBS blew up the 33-year-long oldies format of WCBS-FM, fired the air staff and replaced it with Jack-FM, the biggest programming fiasco in the history of the New York market.

Today - weekend sports update anchor at WFAN New York... and producer and principal reporter for Premiere's Hot Wax Daily, the leading entertainment news and feature service for classic hits radio.

Dick Summer

Great idea Robin.

First radio gig: while I was in school...part time...WNRC in New Rochelle. Worst experience...the night I quit, I was doing the 15 minute sign off newscast, and the receptionist (on whom I had an eye, but no fingers) stood in front of the studio window and did a slow strip. When I hit the weather, she zipped up, blew a kiss...and ran out. Best experience...my first fan letter from "Kathleen."

Worst radio experience - getting fired at WNBC in the "Pittman Purge."

I'm a Brooklyn guy, and I worked at NBC as a page while I was in college. Working at WNBC was a childhood dream.

Getting fired was also one of my funniest radio experiences. Bob was new to New York. He's a Chicago guy. He called me into his office and said, "I've never heard anybody who can communicate on the air like you do. But we're not doing that any more. You're fired. But you can take your picture from the lobby if you want. We won't be needing it."

My current day gig is Communications Director for a national law firm...it includes doing their TV commercials. My "real" gig is www.dicksummer.com. (Signed) Tarzan. ***(RM-Ha. Ha.Jayne reference here! Love you Dick! Please take a listen to this: http://www.box.net/shared/dm9hi7ep28)***

Ron Harper

My first gig was at KWXI (beautiful music) Ft. Worth. 1970. Remember when we all had to have a Third Class license? I got mine, and headed out with a list of the AM stations under 5000 watts, which was what I was licensed to operate. It didn't mater that I lived in Dallas Ft Worth, a top ten market. One of my early stops was at the station I had listened to growing up: 1270 KFJZ. The PD listened to my awful demo tape, and to his credit kept a straight face. He asked me if I had tried their FM station. I didn't know they had an FM station. FM wasn't very big at that time. The PD of the FM actually held his hand behind his ear when he was on the air. We talked for a bit, and then he spelled Andre and asked how I pronounced it. He asked what Andre's last name was I knew Previn and Kostalanez. He spelled Duchin and asked how it was pronounced. I knew it, and I knew the first name was Peter. Apparently listening to my folks' music, and constantly browsing the albums at the dime store wasn't such a bad investment of time - I got the job. Weekends at first, and then as summer progressed, five nights a week. But it was radio and it was cool, and I got to work down the hall from all the guys I had listened to for years.

I'll have to continue this later. Let me think about the "nightmare" part some more. ***(RM- uhh Ron...? If you have to THINK about a nightmare? Don't bother! You're one lucky dude! Borrow ours!)***

Ralph Allen

My first gig was WWNR a 1000 watt AM rock station in Beckley West Virginia. I was in the tenth grade and I immediately went from a frog to a prince. That job bought me a brand new 1965 Mustang.

I went to college early, to the University of Charleston. I worked for WCAW and WVAF (V-100) in Charleston, West Virginia all through my college career. One night, I was on the air, and a maintenance engineer was under the board soldering a wire...and he touched my leg with a hot soldering gun...and I yelled "JESUS CHRIST!!!!" in the middle of a weather blurb.

Well, WCAW was a 10,000 watt MONSTER of a radio station...Clear Channel 680...and it want over parts of two states. Moments later the General Manager kicked the door open and yelled at me, "Godddd Daaamn Ralph...did you just say "Jesus Christ on the air???!!!"

"Uh yes, and you just did too, Paul...the microphone is still on..." He retreated quickly...and waited for me as I left the studio...one of the nine times he fired me and hired me back before I made the 45 minute drive home. I worked for Capitol broadcasting nine times. My GM was a great radio guy, but absolutely way out of control, God rest his soul. He's gone now...but his ghost still haunts me. LOL

I was a morning jock at WVAM as Jim Reese, i.e. Grant Hudson had just left to do news at CKLW. Charlie Fagan, was PD and now a retired vice president of ABC and formerly of Westinghouse, and KDKA's chief of engineering operations.

Charlie worked with us in Parkersburg several times. He tells the story of the transmitter being a 'Continental', that would re-set itself after a power outage.

The interesting thing was, that it would come up a full power...load into the directional antenna system...so you'd be at high power directional after a lightening strike...that knocked the power off...

Also, the microphone switch was a silent mercury switch that also would change value and be "on" after a power outage.

So Jim Reese was on the air, WVAM took a lightening strike, and the power came back on...filaments hot...the transmitter came up to full power, directional antenna system...and just as Jim yelled "Son of a fu@#%-in' Bitch" as the microphone was hot...and it went over seven states...and not ONE phone call...not one... ***(RM- THIS I find to be the most amazing point in the book! 7 states- not ONE call?!)***

Jesus loves Jim Reese...i.e. Grant Hudson.

AFTER he went to CKLW he came back to visit, had a new Chevy Super Sport, and a roll of $100 dollar bills wrapped up in a big shoestring...and he took a magic marker and made all the 8's bigger...the big eight...cklw... remember?
True story, I saw the clocks...it really happened.

AND...

Do you want to hear the cash call jack-off story? LOL That was kuel!

The "Son of a fu@#%-in' Bitch" loaded up on the directional antenna system at full power one morning...broadcast over seven states when I was at WVAM in Altoona?

In those days...the FCC was bad bad bad on ya' too.

I'm now working at a 1000 watt AM radio station in Marietta, Ohio for a family owned trio of radio stations, doing mornings. I had the same job some 25 years ago. They leave me alone, never tell me what to play, what to do, how to do it, or anything but "Glad you're here!" and the paychecks are good and there is one for me every other Friday afternoon. Life is so good. AND I get to shove one up Clear Channel's butt every workday of my life...what could be better than that!!!???

*you may not want to print that...but it's true! ***(RM-it's YOUR story, not mine! Why would I not keep to the truth?xoxox)***

Rock on!!!

I love ya!

Kevin Fitzgerald

"Hi Robin hope you are well.

My nightmare was in college at the University of Missouri-Columbia in 1977. I was the Sunday night host for the Boston Symphony on KBIA and while I had my share of botched names of various artists my nightmare came one night when I put on the tape of the symphony and after awhile I went down the hall for a biology break. When I came back there was tape all over the floor and the door locked behind me. I went into panic mode and ended up knocking down the sound proof door. The chief engineer at the time of course had heard the dead air and was on his way over.

By the time he arrived I had restrung the tape and made my "technical difficulties" announcement and was back on the air. He came into the studio and saw the door complete with its hinges leaning against the wall.

All he could do is laugh, he didn't dare comment.....I had just knocked down a very large, heavy and secure door.

Good luck with the book. Fitz"

Mary Lou Jones

In the mid-90's I was on the air on a Saturday morning on WSTV-AM in Steubenville, Ohio. A former neighbor of mine called and asked me to play something upbeat for her mother, who was 90 years old and recovering from a broken hip, but feeling very low and very old. I thought I cued up "Love Will Keep Us Together" by the Captain and Tennille, but when I hit the remote start on the board, the first note told me I had made a big mistake. Instead of C&T, I started to play Bob Dylan, "Knockin' on Heaven's Door." I was mortified!!! I played with the pot to make it seem like I had trouble with it and grabbed a CD and threw it in the machine. I think I ended up with a Smokey Robinson song.

I called my then-husband and then my PD to tell them about it because I was so upset, and both of them almost dropped the phone because they were laughing so hard. ***(RM- Toooo funny!!!).***

Today I am still with the station (now Froggy 103.5 & ESPN 1340), working as the copywriter.

Michael Marvin

My first paying gig was at WLOU in Louisville. I didn't have any nightmares working there. But one funny incident was I was running a spot for women's hair care product and I cued up a James Brown on the air and not in cue. So you heard a woman talking about great hair, who would be using this product and at the same time ...one would hear "WOOOOHHHHHH AHHHHHHHHHHHHH WOOOOOOOWWW." Kind of funny in a way.

Mark Biviano

Haven't been on the air since the mid '70s.....STILL have the the nightmare every so often. Nothings works....can't find the next record...I'm doing a format I don't really know...the records I DO find...I can't read the labels. When I open the mic...I just utter garble (some might have said I did that in REAL life as a jock).

Scott Blaine

Sometimes I really hate voice tracking...i voice tracked for two stations, one right after the other...First for Buzz 105.7(AC) and then Country 107.7 WGNA in Albany....must have had the Buzz still on my mind for the first break on WGNA cuz I'm in my car listening when I hear myself say Heres' the latest from Glorianna, ...Wild at Heart on Buzz 105.7!!"" DOH!!!

Don Blesse

WRLB Long Branch, NJ (now part of the highly gerrymandered Monmouth-Ocean Arbitron metro #51)... Summer 1972 and was doing the transmitter readings (every half-hour in those days) and noticed that the stereo subcarrier was down and we were broadcasting in mono. The chief engineer was called. When he called me back, he said the fix was simple...

Ever since I started at the station we had been running on an old Gates backup transmitter. The main was in need of some sort of part and the owners were in no big hurry to shell out for repairs. The backup had this annoying little habit where the power supply relay would cut out and the transmitter would shut down. This would happen at least once or twice per shift. Normally a quick sprint down the hall to push the start button would get us back on the air. One Saturday, however, the relay cut out and as soon as you released the button the relay would cut out again.

Apparently we were still broadcasting with flea power. You could hear a hissy signal through the air monitor. The stereo subcarrier exciter was still on (thankfully it wasn't a hot day or that would be down too, as per my previous post) providing a rudimentary carrier. According to the frequency monitor we were 19 kHz off-frequency.... which was quite illegal.

My first call was to our chief engineer. His wife answered the phone and informed me that Charlie was out on a boat fishing and wouldn't be back for six or seven hours. (There were no cell phones or pagers in those days.) My next call was to the program director. No answer. Next, I tried the General Manager and got a busy signal (no call waiting either in those days). After about 30 minutes of bouncing between the PD's no answer and the GM's busy signal, I dialed the Operator and convinced her it was an emergency and to break in on the call on the GM's line. His wife came on the line and she told me that he couldn't come to the phone at the moment. I explained the situation but she said her husband had just had oral surgery and was still sleeping off the after effects of the anesthetics and painkillers. I asked her to please have him call me back as soon as possible and tried the PD's line again, to no avail.

Another half-hour goes by and the Gm finally calls back, sounding still a bit loopy from the drugs. I explained the situation and when he heard that we could be heard, albeit for no more than a few hundred feet he said, "Well I guess we're technically on the air," and told me to resume format, run and log all the spots including the ones we missed earlier, and that if any clients complained he'd give them make-goods. So, for the next few hours until another engineer could be located, I was broadcasting a noisy signal to about a three-block radius around our tower. ***(RM- Don? You have just proved my point! I have no idea, what in the hell you were just talking about! THUS- engineers run when they see me comin'!)***

John Russo

It was 1966. I was never very good with the complicated system at the station. I was supposed to patch in a remote. Instead, I found myself airing my boss' private conversation with his girlfriend!

Michael Filipelli

WWTR Bethany Beach, Del 96 Rock…

I was pulling an airshift on the sister oldies am station WETT. Flat roof on the building was always leaking and forever being repaired. It got sooo bad we had to staple a clear plastic cover to the joists in the ceiling to catch the water and puncture a hole in the plastic so the water could run through the little hole and into the huge trashcan behind the airchair. Yes, WETT-am!

Paul Buck Bokuniewicz

Went to bathroom...door to suite locked behind me. 2am...trying to communicate with the only other person in the building...a deaf guy in the suite down the hall. He gave me a boost as I removed a ceiling tile and went over the door jam. Lots of dead air.

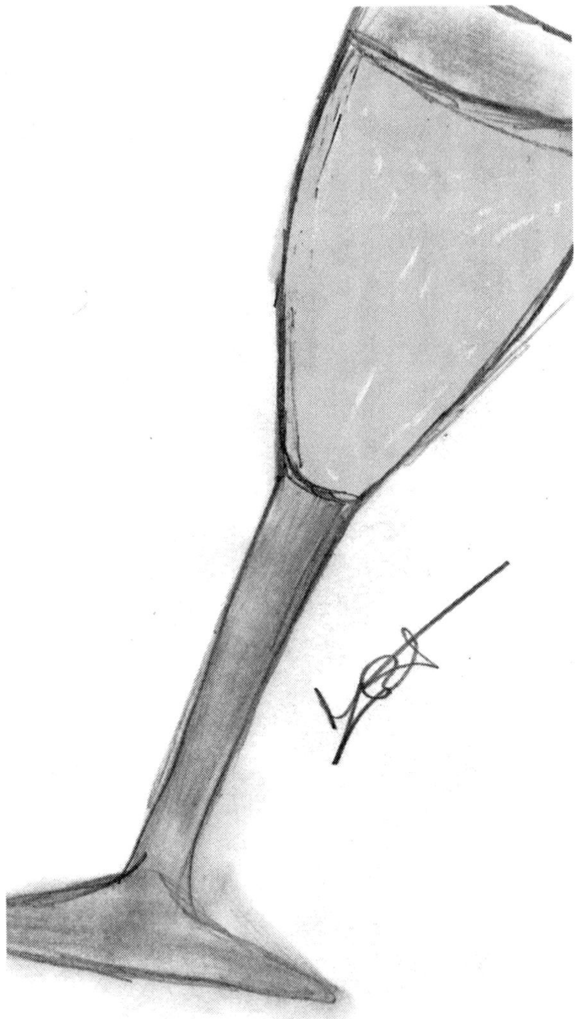

Mike McCay

Just outside Washiington DC, a Wine Festival is held in an area called Virginia Hunt Country. Be very careful when giving the location. Very careful. It doesn't look good on your resume.

AND...

I remember a local sports guy reporting from a Pro BBall playoff championship post game party saying. ' Yesss, they're beginning to sh*t the sampaign, I mean sip the Champaigne"...I have that aircheck somewhere. :-)

Gary Russell

I was doin' overnights on WKLC, an AOR station just outside Charleston WV back in '81. 'Bout 3 AM I gotta' request for "ANYTHING" by Led Zeppelin, so I obliged. About halfway thru the song, I reached across the turntable to cue up the next tune. Even though this was June, the station was always freezin' overnights, so I wore a light, long-sleeved shirt. The shirt sleeve button had gotten lost in the wash, and the dangling cuff caught the tone arm playin' Zep. ZZZZZZZZZZZZZZZZZ!!!!!!!!!!!!! I hurridly dropped the new record on and fired it up! Phone rings immediately, turned out to be the listener who'd requested the Zep in the first place. "Duuuuuuuude, what was THAT?! We'd all just got here from the bar when I called and requested the Zep, and thanks for playin' it, but...I live way out in the country, ...and had it crankin' FULL BLAST thru my 500 watt stereo, then there was this loud noise, and I blew two woofers and a tweeter!!! Wha' happened??" Tryin' not to laugh, I blamed it on the owners "cat" who'd jumped up and hit the tonearm with his tail. "KILL THAT DAMN @#$%^*& CAT!" was all he said, and hung up!

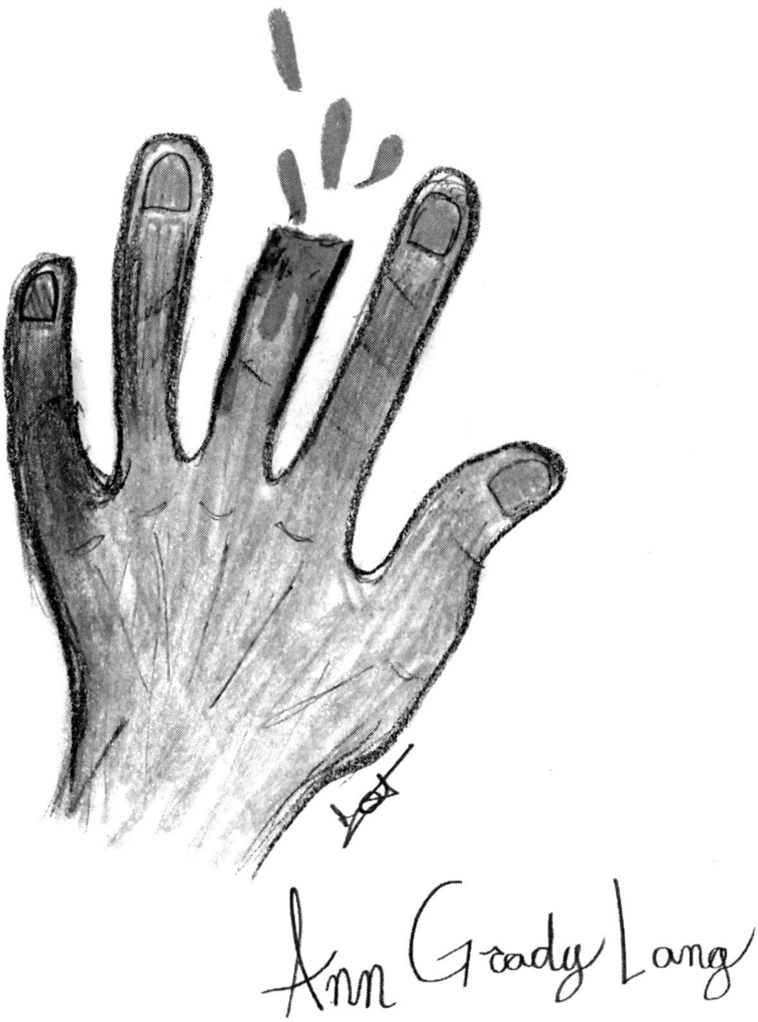

Ann Grady Lang

I am, quite possibly, the only person who has amputated a finger at a radio station. 17 years ago; 400-lb soundproof door to the studio; blood spurts still on the wall last time I checked. Middle finger, too...go figure! ***(RM- it's almost poetic justice, in our industry, if you held that finger up today?)***

Bill Cain

It was 1973. I was still a Junior in High School, and determined to crack into broadcasting, rip, shit or bust. I hung around the town's lone station, a day timer at 1370, WPAZ, in hopes of learning and picking up part-time work. One Saturday, a jock didn't show for a remote that the owner, Herb Scott, was hosting at a breakfast downtown for charity. I volunteered to set up a banner, wire the mikes, hook up the mixer and ride levels. When Mr. Scott arrived, it was all set, and the lines were tested the WPAZ way...'clip the audio if it sounds good, clip it twice if too hot'. Fine.

Now Herb was somewhat short in stature, but tall in business in town. He was a force. He also had pronounced facial features telling of his Middle Eastern decent and religious heritage.

He dons a canary yellow blazer with the call letters embroidered in red and begins to work the room. The "room" consisting mostly of old women with shopping bags getting pancakes and sausage for a donation.

With wit, warmth and charm, he strolls the tables, and in his best radio voice begins interviewing the patrons (mostly with their mouth full). At one point he reaches the 'woman'. This woman wanted no part. He asked her a question, and she defiantly put more food in her mouth. Now, Herb was not one to walk away from a 'no sale.' He got her to say that she has been a listener for 20 years. Now Herb had her. He got her to say she loved his voice on the radio. Herb was delighted, but I had other feelings. Then out of the blue she adds, 'But your voice doesn't match your face.' I'm dying of fear of what's next, and I was right.

Mr. Scott replies "you mean I'm more handsome!" ...woman: 'No.'

Herb: "You thought I looked more like Cary Grant"...woman : 'No.'

Herb: "You thought I looked like Robert Redford"... the woman.

THOROUGHLY annoyed, grabbing her bags with one hand and the microphone with the other states with Broadway Stage clarity: "You look like the Jew I always thought you were," and walks away. Herb, crimson red but undaunted..." there you have it ladies and gentlemen, another WPAZ listener with good taste." All of this live over the air. ***(RM-I allowed, because I'm a Jew too! My mom would have called her a 'simpleton.')***

When I got back to the station, the staff had the interview racked on the Ampex machines running dubs for God knows who or what, and a cart with the line playing in cue over and over while they were in hysterics.

Gary "GAZZA" Sands

Started in the industry at SKY FM (Community) which became COAST FM in South Australia after a dispute by Sky channel sport group, they wanted the name in S.A. and decide because we were only community radio we didn't count, how wrong they were it cost them big time to get the station to change it's name.

After Training from the SABN course under the direction of Vaughan Harvey and moving around in the Community sector as announcer and PD/Station Manager, at WOW FM and others I became a lecturer for media studies (Radio) for colleges then moved to Charleville outback QLD doing breakfast I then moved on to Commercial Radio.

My Commercial intro was at 4ZR Roma for the DMG group with the on air call of Gazza for breakfast on Roma and the Maranoa's 4ZR, then off to 4HI after the Macquarie group purchased the stations with a call of Gazza for breakfast on the central highlands and coalfields 4HI, now back in Roma on sale of stations again to Smart radio group where a change came with a new call sign ZincZR, call is the morning ZOO with Gazza on ZINC ZR.

My worst radio moment was when I was introduced to live radio in a mid dawn slot and the guy who was supposed to stay with me decided that it would all work ok even if he wasn't there so off he went, we had 3 turntables in those days and did the big intro sounding good only to find when I turned the table on the needle had fluff all over it and off it went across the record with a big screech and not having anything cued it was a bugga of a lesson, one that I would never forget.

Sandy Weaver Carman

First station: WDAK, Columbus, GA in 1977

It was the fall that the Stocking Strangler started terrorizing old women in Columbus. He'd raped and killed a few, and every woman in town, young or old, was a bit on-edge, and the men in their lives were, too. I wasn't even on the air yet, just was in charge of running the Sunday morning public affairs programming from 6am to noon.

Around 10:20, I went downstairs to get a Coke from the machine and the door at the top of the stairs closed behind me...and locked! There were tools in the basement, but the hinges were on the other side of the door so hinge removal wasn't an option. I tried kicking the door, but figured I'd break a leg before ever breaking through the solid door. The tiny windows weren't an option – far too small to crawl through if I broke one out. Then I spotted the alarm wires. I figured if I triggered the alarm, someone would come to check on the building.

It was now 10:35. The network show that had been on ended at 10:30, then there was 20 seconds of silence, then the network 10 second tone, preparing for the 10:30 program I wasn't scheduled to broadcast.

That all happened 5 minutes ago. Oh, well...at least something was on the air. I could hear the hotline ringing and the calls on the two-way from our sister station, trying to get someone to answer and get back on schedule. I gritted my teeth against what I thought would be a loud alarm, closed my eyes and cut the alarm wire.

Nothing happened. Instead of an alarm, which I knew would bring the police to investigate; there was mind-numbing silence. I paced the basement, searching for a solution or a way out of the nightmare that had turned real. The hotline continued to ring, and the two-way chatter continued. The Coke got warm.

The police showed up at the same time as our Chief Engineer, who had been out on the river, fishing while listening to the station. He knew something was wrong when he heard the ten-second tone, so headed for the station right away, stopping to call police on the way. His only thought was that the Stocking Strangler had somehow gotten to the teenage girl alone at his radio station.

Bless Charlie's heart...the look of relief when he found me, frantic, in the basement is one of the sweetest radio memories I have. I think I just dashed past him, played the legal ID and joined the 11am church service, already in progress, at 11:08. He laughed at me as I told him everything I tried, or thought to try, and he promised that I wasn't in trouble for cutting wires. He stuck around until the afternoon guy came in to do his shift, then went back to fishing.

The lock did get taken off the stairway door the next day, though! ***(RM-Sandy, you have successfully yanked our 'heart strings' with this one!)***

Also at WDAK, I think in early 1978, we were exceptionally short-staffed. There were only three people to cover 24 hours of live programming, meaning that we each were pulling 8-hour shifts. At the beginning of mine, I racked a 10-inch metal reel of tape, so I could capture the network newscasts and play them back on our schedule. Of course the tape needed to be rewound, so I did, watching the second-hand on the clock get closer and closer to the time I had to catch the first feed. I stopped the tape and put the heel of my hand on the reel hub, hoping to slow it quickly so I could record all of the news, which was only seconds away from airing.

Didn't pay attention to where my fingertips were. Heard the blow, and felt a thud on the end of my left index finger. Saw the blood. Hit play + record, and went to the bathroom to assess the damage. Bad. I could see the end of my finger bone, and what was left of the fingertip was attached by just a little bit of skin. I grabbed a bunch of paper towels, put the fingertip back together, and wrapped it tightly in paper towels and splicing tape, because the first aid kit was in pretty sad shape. Thankfully, the finger didn't hurt much until about 7 hours later...Mother Nature kindly kept it down to a dull throb until then.

The doctor in the ER chewed me out for waiting so long to come in, disinfected it and sent me on my way with a sporty new bandage made of actual bandage materials. He couldn't sew it up because I was several hours beyond being able to do that. He didn't understand why I hadn't come in hours before.

Face it...he didn't understand radio.

Recurring Radio Dreams

#1, and by far the most common one – I'm in the studio and everything is so big! The song is nearing the end, and I can't find a chair to stand on to try to reach the board, which is far above my head. Find a ladder, prop it up, but I'm still too small to do anything, because my arms are too short to reach the faders. This is almost an Alice In Wonderland type of nightmare – very bizarre perspectives, and the dream always ends before the song does.

#2, an occasional nocturnal interlude – I'm talking, and the wrong call letters keep coming out of my mouth. I know what I'm supposed to be saying, but my mouth won't form the right words. Instead, it says the wrong calls and introduces the wrong kind of music. This dream seems to last only seconds, but replays over and over on the nights it shows up.

Roger Morgan

My first radio experience may have been my worst. It happened in my hometown of Salem, Oregon when I was in Jr. High School. I was 15 and eager to have my own car. I asked my parents if they'd help and they pointed to my bicycle and told me that 15 year-olds needed to EARN their cars!

The bicycle ride took me to a local radio station, (K-GAY - it was 1957 and "gay" meant "happy". I immediately applied for an on-air position but was told that those guys had to start at the bottom which, in my case, meant "gopher". My first assignment was burning teletype paper in a large field behind the station. (Plus the occasional task of cleaning sinks and toilets...who'd have ever guessed that the "big radio stars" started out this way?!) One day when I returned from burning trash, I walked back into the station to discover that the entire staff had been fired!

The eccentric owner, seeing me for the first time, asked if I knew anything about being a "radio announcer". A quick glance around the place told me that there was no one there to dispute what I was about to tell him and

I proudly said, "Yes, I know just about everything there is about being on the air." In reality, I'd never been allowed in the control room! I was hired! My first day on the job and I was left alone to "become a radio star."

It took me awhile to figure out that there were two turntables. While a record played on one, the other one was set up .... well, you get it. 3 hours into the "program" I decided to actually say something on the air. "A news cast..that's what I'll do!" (Afterall, I had a good deal of experience dealing with old teletype paper.) I found a "five minute summary" of news on the teletype machine. I wondered what words to say as I began the newscast. "Keep it simple"....a good lesson for anyone in broadcasting, and I decided to open the newscast with words I'd heard many times on "big time radio". "...and now, news headlines and highlights." In my mind, it seemed easy enough. But, when I opened the mic and stark terror replaced common sense, the words came out different somehow. "AND NOW, NEWS HIGHLINES AND HEADLIGHTS!"

All these years later, I still recall the perspiration beads that began to form on my forehead as I repeated the phrase over and over, at least 5 times..."AND NOW, NEWS HIGHLINES AND HEADLIGHTS!" Who would ever have guessed that such a disastrous beginning would result in a career that has afforded me so many wonderful opportunities in life. ***(RM- notice we have the same initials? What a WONDERFUL story! Way ta' break into the biz!)***

Today, I'm in the process of selling two radio stations I've been able to purchase and I have relocated from our home of the past 9 years in the Virgin Islands, to new "digs" in Los Angeles. I'm preparing to launch a syndicated talk program from here. Who knows where the adventure will take me? I can't be certain, but as has been the case with every experience in broadcasting over the past 50-plus years, it won't be boring!

Greg Breeden

Somewhere back in the mid-1980s I was working for an AC station that was still running local high school football games. I was the board-op on this particular friday night and all was going well when for some reason (can't remember exactly why) there was a delay in the kickoff of the game so the guys on the field thre...w it back to me in the studio to play some music to kill time. So, being slightly un-prepared for this I reached for the first music cart I could get my hand on, shoved it into the player, potted it up, hit the 'start' button and the song started with the vocals, "this has got to be the saddest day of my life..." Yeah, what better way to add to the energy of a high school football game on a Friday night with "Kiss And Say Goodbye" by The Manhattans! Even after 25 years, I still think about this and laugh every time I hear that song!

Brad Lovett

I've always had the Dead Air Dream and still do to this day. Usually I'm in a studio with the mike open, nothing cued and I can't reach anything. Sometimes I'm at a station playing vinyl. The Dead Air Dream became real once.

I was working at the late "Kool 95" located outside of Dayton, OH in Piqua on a Saturday. The jock before me locked the back door and set the alarm. Then the 85-year-old owner saunters in the front door, setting off the alarm, and the code the PD gave me didn't work! So here are sirens blaring, the phone ringing, Police showing up and finally the GM. To top it off, the automation computer picks that very minute to freeze up and need a reboot! There was nothing even in the building to put on the air. Figured that would be my last shift but all was OK and I spent a little over a year there.

Ron Chavis

I stumbled into professional radio (WDVE), submitting a demo tape only after much prodding from some pals at the Point Park College (Pittsburgh) station where I spent much time smoking grass and reading Carl Jung during long Hendrix records (been writing too-long sentences ever since). These were the days before "the music died."

The Feds had mandated that the Big Three in Detroit begin installing FM radios as standard equipment in all new autos, and the big radio companies who'd been busy gobbling FM frequencies while lobbying for that mandate were now scrambling to find the best niche for their new stereo broadcast capabilities. What they found was "Album Rock," the mix of everything from Uriah Heap to Carole King to Led Zep that was making waves through WPLJ in New York. In Pittsburgh, WDVE management was looking for announcers who knew that wide array of music, knew the peculiarities of Pittsburgh...what the city wanted to hear, musically and otherwise. We were all in uncharted waters here. So, the false science of "programming" (spoon-feeding the audience with music and jargon according to the dictates of so-called market research) hadn't yet contaminated FM to the extent that it had AM Top 40 at the time.

Believe it or not, most FM deejays actually had some say in the music we played in those fledgling days of "FM." Indeed, we were mostly college kids, or of that age, who had some radio experience and who happened to be in touch with what our peers wanted to hear. We were craftspeople, though I doubt any of saw it that way; and we were truly the best at what we did. In 1970, our little staff of radio "nonprofessionals" at the newly staffed WDVE wound up beating the headphones off of our "professionally staffed," tightly programmed AM sister station, the Top 40 KQV, in the first ratings book after our staff hit the air. Talk about instant fame... and jealousy... and the flames it brings: the seasoned, "ace" announcers from KQV were suddenly not speaking to us newcomers, and even demanding their own bathroom so as to encounter us as little as possible.

FM rocked to the top in LA and Boston, Houston, even Toronto. You KNOW the part FM played in that very special era, if you lived there. It made mags like Rolling Stone happen. It made stereo and head shops boom industries. It made my career happen.

And everywhere that this "Freeform" sounding music and personality style of radio prospered, the competition envied it... and competition everywhere tried to replicate it by researching it... and creating new programming to implement deejay compliance to this new "format" or that. Until, in a few short years, FM had become what it had replaced.

Inevitably, sadly, laughably, FM became the blind, ugly giant that AM Top 40 was in the first place: a barren wasteland of repetition that cheats everyone --

from music performers who don't fit the "direction things are going," to the poor, pablum fed listeners whose minds are probed via telephone surveys, Focus Groups, "listener song testing" and "buying trends." Not that these methodologies and resultant formats are inherently bad; but they are often completely useless tools in the hands of white shirts who have more of a background in statistics than they do entertainment. If Prince were to have begun today instead of yesterday, there's no way you'd hear him on Radio: UN-trendy! And, almost universally, with the exception of clown morning show hosts, deejays' lips are kept in closer reign than Toni Braxton's bra straps: A fact that has allowed for an abundance of “communicators” with zero talent and nothing but a driven desire to be “on the radio” to fill the “talent pool.” It’s ironic that today, smart-minded, “people meter” conscious programmers complain of a difficulty in finding real talent who are capable of “entertaining” in a way that paints verbal pictures for, and makes the subject matter at hand ABOUT, the Listener.

A few Radio Artists I hear today are capable of that. And they stand out like surviving pyramids against a skyline of McDonald's "golden arches" and red and white TGI Friday awnings... shining testimonies that the basis of true craftsmanship doesn't lie in some ever-evolving formula, but rather a tried and true knowledge of the material at hand and the skill to work with it. And that is artistry. That is craftsmanship. That, like it or not, is the real crowd pleaser, the sine qua non of genuine entertainment. That’s my take.

Today, I’m doing 13 shows a week. I’m performing an R&B Love Song show at KRNB 105.7 FM (Dallas/Fort Worth), Monday through Saturday evenings, 7PM – Midnight, Central Time. It streams live, via www.krnb.com. And another program, very similar in nature, that’s syndicated via Citadel Broadcasting (to 49 cities currently). A live stream is here http://www.thetouch953.com/ for the show, which runs 7 nights a week, 6pm to midnight, Central Time.

Ron Chavis ***(RM- One of my favorite ‘burghers.’)***

Aka, Rockin’ Ron

Aka, The Nighttime Dog

Jason Lord

Hi Robin,

My first gig aside from Student Radio, was a sports update anchor/board-op for espn radio 1450 in State College.

The worst radio nightmare is a tough one. During a sports update early in my career. I wrapped up my 2 minute update and then as the producer tried to tell the host he had a spot coming up. The problem with that was my microphone was on so I not only told the host about the live update coming but I also told all the listeners.

The close second was when I worked as a high school play by play broadcaster. First my van has broken down earlier in the day, so I rushed to take a cab to the game by 2pm thinking it was a 3 o'clock game. It was a 4 o'clock game which was mistake number 1. Then as the game would down I emphasized the point that State College has to look at the play clock as they run the time down. The problem was as it came out of my mouth I missed the L and said" State College has to look at their play cock as they run the time down." Thankfully the guy doing the game with me and the assistant in the studio were the only two people to catch it and very few listeners noticed. Those were likely my worst nightmares aside from the obvious occasional empty break or mind going blank etc.

I'm Still in State College working as a weekend FJ on 93-7 the bus 12-4 and still doing everything for the sports station espn radio 1450. For 1450 I cohost a weekly show and do play by play for high school football and basketball. I also work as a radio broadcaster for Accu Weather in State College.

Hope this answers your questions on first gig,nightmare and where I am today. Looking forward to the book:)

Jason

Joel Dearing

First job in radio:

Weekend announcer at WLAR in Athens, TN.

At that hole in the wall station the front door of the building could be locked from the inside, thus making your key from the outside useless. Why, I do not know, that was the way it was. The guy that was on before me was a real practical joker, and he happened to be at the station one Saturday night. After all what is there to do in Athens TN on a Saturday night?

At the bottom of the hour we ran Mutual network news. It was a nice five minute break in the hour to stretch your legs, pee, and get a breath of fresh air. I did mention that this place was a hole, and a very small hole at that, and everyone on the staff smoked, except me. So the building reeked of stale cigarette smoke. This was a time when you could still smoke in the building and studio. Taking my five minute break, I walked the 15 steps to exit the front door, take a big deep breath of clean east Tennessee air, stretch and suddenly I hear the door slam behind me and the hard clank of the interior lock. Peering through the windows I see "my friend" go into the studio, remove the tone arms on the records I had cued up, put the records back in the shuck, and back in the rack, removed the jingle from the cart machine and turned off the cart machine. He then turned off the light and closed the studio door.

Being new in my career and NEVER wanting to make a mistake, I am going ape shit (can we swear in your book?) As the seconds clicked by of my five minute window, I see the biggest disaster of my young career, and the PD lighting up the hot line wanting to know why I was asleep at the switch, and to just leave the key on the desk when I am done, and maybe that job in the factory will still be there for me, and I can go back to chrome plating stove parts for the rest of my life.

At 4:50 into the Mutual news with Cami McCormick "my fried" unlocked the door. With a few choice words about his heritage, and parentage, I flung myself into the studio, grabbed any record as Cami's alto voice said "This is the Mutual Broadcasting System be dupe" (only the really old, old radio dogs will get the "be dupe."

The record I grabbed was Earth Wind and Fire "Sing A Song." In my rush I plopped it down, no slammed it down on the turntable. I dropped the needle on it and hoped that the turned table was still on the 45 speed. What I did not know was that the promo copy of that song was not the same on both sides like most promo copies are. That record had a B side. The B side of the promo copy of Earth Wind and Fire "Sing A Song" was an instrumental version of the song. So as I am trying to compose, I am thinking... "where are the vocals?" "Why is this different?" "Ah crap, maybe I can sell shoes, is there a big future in fast food?" I wish I wasn't so scrawny, and my friend wasn't so much bigger than I was because I'd sure liked to decked him."

Today I am V.P. of Affiliate Relations for Virtual News Center. VNC is a company that out sources local news for station across America. As of this writing we are in 35 markets and have about 70 stations using our service. ***(RM-Even though brief, it was great working with you Joel! And THANK GOD we did NOT move to Green Bay!)***

Mike Belancsik

1st gig was not half bad, worked at WREN, ***(RM-Weird here...that's my eldest daughter's first name! WREN)*** a clear channel west station in Topeka, Kansas. They played middle of road stuff such as Frank Sinatra, Tony Bennett and Ella Fitzgerald and with my erotic voice it worked!!

My boss decided my name would be Michael St. John and my first show at 1am was called "Music for lovers and werewolves," stuff like "I only have eyes for you" and "Left my heart in San Francisco." I would sneak in tunes like, "The warmth of summer" by the Beach Boys and "What a day for a daydream" by the Loving Spoonful. It was a popular show with lots of interaction and lots of spilling your romantic guts!!

My following went past the normal romantic chit chat, we had those gay blades who needed a voice and because it was before disco I was picked and was asked more times to breakfast than Captain Crunch!! So I'm not gay but it's still nice to be popular!! Maybe I have good taste. I do like pink and chartreuse.

I was on the air when we first walked on the moon, still have that tape and the tape of the shooting of Robert Kennedy.

My worst experience was a stomach flu night when I became up and close with the toilet. Played back to back songs of "MacArthur Park" to give me the time to make it to the throne.

Also I had my first girl and boy stalker, the girl wouldn't take no for an answer and my wife would get phone calls in the middle of the night saying she should leave me because I was seeing the stalker. Think about that, my wife didn't know for sure, finally got a restraining order!

The president of the company ran for president of the United States in the 30's and used to call me late at night just to talk, I guess he had insomnia, he was a nice man.

Got a job offer to go to Chicago and always thought this president put in the good word!!

Bill Figenshu

It was WMID (1340- The Jersey Giant) in Atlantic City NJ. $80 a week and all the 45's I could take home. My shift was Midnight to 6AM Monday through Friday and Midnight to 9am on Sunday morning. 6 to 9am Sunday I had to run the board for the gospel show that was broadcast LIVE with a full choir and band. (Choir robes and everything.) To this day I have a special love and appreciation for gospel music and the passion it creates for the listener.

I was 19 years old and lived in a rooming house with a few other jox. The rent was $75 a month with hot and cold running cockroaches, yet

I considered it one of the most exciting times in my life. I was actually getting paid for doing what I love. By the way, 40 years later (I started June 1970), there are days I hate loving radio as much as I do, yet, there is still the passion for this medium I wouldn't trade for the world. Radio people have a bond with listeners, advertisers, and each other that few understand outside of the medium.

Stupidest career moment: Hmmm, it's hard to pick from so many, but the dumbest would be climbing WMID's 200 foot self supporting tower with another jock. It was cut off at the top to allow for a 5 foot square platform. We drank a six pack of beer while enjoying the view from the top. Young and stupid? Oh yea!

Worst career moment: leaving a high paying corporate job to go out on my own and reinvent myself. It's very scary to walk away from something you have done for so long but I needed to immerse myself in learning about new media platforms. Creative experience means nothing if you can't apply it to changing technology.

Best career moment: leaving a high paying corporate job to go out on my own and reinvent myself. I am fortunate to have wonderful clients that allow me to be independent in thought while earning a great living. Creative experience is an advantage when applied to changing technology.

Robin, Is this what you want? Is it ok? ***(RM- "HE" is asking ME, if this is 'ok!' THIS is why I love him.)***

Bill Figenshu – President, Figmedia1

Matt Steinberg

Since you are going to a second printing, here's my story, finally!

First on air job: Weekends and fill-ins at WGLI-AM/Babylon NY. I started out as an intern under PD Dennis "Dick" Moore and learned more from him than I ever did in any communications arts class. How many PD's take the time to actually show someone how to wind carts and let you be creative anymore? I did the usual intern stuff as well, production and even writing spots. I also did person on the street interviews at the LIRR train station, and a local strip mall, asking questions relating to events of the week. I got to produce this myself every week, and they got played during the broadcast day for the entire week.

The highlight of my time there was interviewing, and riding on the press truck when the New York Islanders won their first NHL Stanley Cup during the 1979-1980 season. I got to interview most of the players, and had those play for weeks afterwards on WGLI.

Worst on air moment: I was hired by Larry Miller, former PD of WKTU-FM and WQIV-FM to do weekends and fill-in's at WWHB-FM in Hampton Bays. My interview consisted of riding around in Larry's Jeep smoking pot and talking radio and music.

He hired me, and I quite my job for retail eyeglass chain. I started two weeks later, and moved into a rooming house in Riverhead, about 10 minutes away from the station. I was put to work doing the traffic logs, which were done by hand on a typewriter, as computers were not around then. Needless to say mistakes were made, and the salesmen weren't to happy when a spot was left off the log, or duplicates were put in by mistake.

After 3 days of being at the station, the over night jock hurt his back, and couldn't come to work. Larry told me I was going to do that shift, in addition to my daytime work. Two problems presented themselves at the same time. The first was that I hadn't been on the air yet, and had no clue what to do, as Larry was not what I would call a great teacher. (That's putting it mildly!) Second, the room I was renting was very bright during the day, and so trying to get some sleep after getting off the air at 6 AM was difficult as I am not a daytime sleeper.

So put no sleep, and total lack of guidance together, and disaster strikes during week two of working at WWHB. I was dead tired on the particular Thursday morning, and missed hitting the news at the top of the hour by a couple of minutes. Larry told me to play another song and just fade it out when the news came on. Ok, easy enough solution.

The next day, it happens again so I put another song on and fade it out to the news. Larry comes running into the studio, sweeps everything off the side counter and yells, "I'm sick and tired of this shit! You're fired! Mail me a W2 and I'll pay you then!"

I leave the station in shock, and call my mother and tell her I'm coming home and I'll explain when I see her on Sunday. I figured I may as well enjoy the weekend in Hampton Bays and hit the beach. The next call was after 9 AM to my former employer, who was thrilled to hear from me, and was happy to take me back, as the guy they hired to replace me wasn't working out. Stupid me, I didn't ask for a raise.

Today, I'm working in retail sales, and running www.bigappleairchecks.com and writing a weekly column for www.grubstreet.ca.

That's my story, and I'm sticking to it! Matt "Magic Matt Craig" Seinberg.

Robin Marshall (My Turn!)

How did I get into radio?

I LIED MY ASS OFF! That's how. I was a rock n' roll singer, songwriter, and the front person of my band called Entourage. (we used to play in front of the Good Rats! LOL!) We played all over NYC, including TRAX and CBGB's, out on the Island, US Blues, My Father's Place, etc... We had a record and were trying desperately to get distribution on it. Had it played on WBAB's Homegrown show too! Haha! What memories...

We came up with the idea, that because I had an 'ok' voice, I should try and get a job on the air, so that we could sneak our record in!

I had my 'pre husband' write me up a fake resume on being an air talent, because years ago, HE had been on the air, and knew a bit about it. He wrote that I'd been a PD, MD, on air… I didn't know what a PD was! I then went to Bob Shannon- yes, CBS FM's Bob Shannon. He was from Pgh, and so was I, and it seems that people from Pgh will do anything for another Pittsburgher… so he brought me up to his studio, and helped me to make a FAKE AIR CHECK! It was representative of B94 in Pgh! LOL!

Now! I was ready to do this! When you're young, you think you CAN do this kind of stuff!

I walked in to WAPP right here in NYC with both my resume, air check and short skirt, the day Mark Coppola walked out. Steve Ellis, PD, was stuck and needed someone right away, hired me on the spot! That night, while on the air for the first time EVER, I called my pre-husband, and he walked me through how to use the pots, (until then, I thought pots were solely for cooking) talk into the mike, thread a reel to reel machine, OMG… when I think back, I was out of my mind.

Flash ahead 6 months, that's when I told them I was a singer, and had a record out. Needed to build credibility first, right? Well, they LIKED it! It was put on steady rotation for a while, but still no distribution. So that was the end of the band, but I stayed on the air!

Who else do you know that can say they STARTED on air in NYC, and STAYED on air in NYC for 25 years?

Moral of the story: (you know by now, I always have a moral) NEVER SAY "NEVER." TEACH YOUR KIDS TOO!!!

You can be anything you want to be, when you grow up, as long as you think 'outside the box!'

My most horrific story, although I've got MORE than a few… 2 really come to mind, and they were both at KJOY on Long Island. After being on the air for so many years, in the city, I decided while having my babies, I wanted to be closer to home for a while, so to the Island I went!

First story, I was to be interviewed by the PD at KJOY. Showed up with my cassette, LOL… only to find out that he was fired the day before! No one called to tell me, and they had no PD to interview me so they sent me to Dean Anthony's office, who was running WHLI!

I sat across the desk from him, him being a legend, me being Wren and Dylan's mom, and then a radio personality... always had my kids come first...and I handed him my cassette. I was a nervous wreck, for a change! We were both uncomfortable, being thrown into a horrible situation, trying to make the best of it. I'm thinking to myself, "Ugh...he's going through the motions, there's no way in HELL he's hiring me! He wasn't even prepared to MEET me, let alone bring someone new into a station he doesn't even RUN?"

He clicks the PLAY button to hear my air check. What happened next was a huge piece of 'history' in my life. We heard through the speakers, "Can Wren say...PEE PEE?" You want to talk about 'DEAD AIR!' OMG. I'd brought the wrong tape! This tape was of me teaching my 1 year old to talk! At the end of the VERY pregnant pause, when we both had realized what I'd done, Dean BURST into laughter, so much so, that he was crying! I was still a bit in shock... but the next words out of his mouth were, "You're HIRED!"

Horror #2 was inclusive of that same child- Wren. I was on the air, the engineers were working on the back of the console which meant they had the plate off.

My husband was supposed to stop by and drop something off for me. In he walks with my now 3 year old at the time, big smile on her face... she looks at the open area that happened to have colored buttons on it. Before we could stop her, she shrieked, " Oooooh Mommy! Pretty button!" With that, she pushed one, and shut the whole God Damn station down! PD- who 15 years later, became my partner with Jayne, (who knew?) came racing down the hall- GM following quickly at his heels- I said to my husband, "Pick her up, and get out now!" HE DID! 'HOLY TOLEDO' did the 'you know WHAT', HIT the fan!!! Took us 20 minutes to get the station back on the air!

Now you know why I have a fear of engineers! LOL!

I find it fitting to end this book with a story from my husband Steve. He too, as I'd mentioned earlier had been on the air for years... take it away!

Steve Bauver (the husband)

I have known Robin for over 25 years and have been married to her for the last 21. Believe me, I could write my own book on all the situations dealing with her, just my chiropractic bills alone from shaking my head in disbelief and wonderment regarding her everyday life of happenings and shenanigans, could have helped fund just one of our kid's college education. However, the one situation I wish to share with her readers combines her being 'her' and the subject of radio.

I was on the air back in the early to mid seventies, you know, back when radio was still fun and you actually enjoyed going into the studio never knowing what was going to happen each night. Well, one of the part time jocks, his name was Dan, had a little girl, maybe five years old. I used to tell Robin this one story how Dan, who was divorced, would bring his daughter, Catrina, to the studio at night, when I was on the air. He would go into the production studio and Catrina would have the run of the station! I was doing nights and only myself and my newsman were ever there. I told Robin how cute this little girl was, blond hair and big brown eyes. She was one of the reasons I wanted kids. Dan used to tell me she would always ask him "when are we going to Steve's house," in her young mind she always thought the station was my house and he never told her differently. I told Robin all about this little girl and her father, and how enjoyable they were.

Let me now fast forward to 20 years later, I am now with Robin and my old station has a reunion/picnic. Dan has remarried, just recently, and I introduced him to Robin. Being the bubbly person that she is, Robin says to Dan, "I have heard so much about you, it's a pleasure to finally meet you!" She then turned to his new young wife and says, "And you must be the daughter I have heard so much about." Gave new meaning to the

term "Dead Air."
Just so you know, I am still shaking my head, but I now have the company of my three older ones to shake theirs right along with me. ***(RM- paleeeze! Anyone would have made that same mistake! Besides, you may THINK you have the three older ones shaking their heads at me as your allies, but in reality, all five kids are JUST like ME. They're just in a bit of denial, is all. Good luck with THAT, Mr. Husband!)***

Hope you enjoy her, as much as I do!

Steve

EPILOGUE

Thank you all so much for taking part in this wonderful book!

As you can see...we'll need to have a second edition, because too many of you were too SLOW... !!! Now that you see what great fun it is, please join in.

Robin.MarshallsOffice@gmail.com

Send me your three stories today!

1. Your first gig
2. Your worst radio disaster!
3. Where you are today!

I'd also like to let you know that my NYC Radio, 'Tell All,' book will be done soon!

This is 'the dish!' The Main Course!

You'll read about "who did this,... that... and the other thing, to who and HOW!" All the well known radio people that we know and love!

Thank you for participating and I hope you had as much fun reading about each other as I had, fixin' you up!
We are a 'motley crew,' for sure!

Robin
xoxoxox

If you need another copy as a gift for someone, order from the below website!
www.RobinMarshall.com/book.html

When you look at the last page... you will understand *WHY*, I had my children.

My son drew this for me, because he wanted to dedicate something *just* for *me*. It means more to me, than I believe he realizes, not only because I didn't ASK him to do it,...but because the picture signifies my roots here in NYC, the trauma of what's happened in our world, but also, the love and faith and hope that is still alive within our hearts- because of radio.

My family may very well be leaving NYC soon- and it's very difficult to just "pull up those roots," and go, but he sure did draw what I feel. It's that "Empire State of Mind."

Love to you, for taking part!

All the best to you- from the bottom of my heart.

Robin

PS! Stay in touch! Robin@RobinMarshall.com

mom

Breinigsville, PA USA
21 September 2010

245850BV00004B/21/P